建筑立场系列丛书 No.64

探索瑞士建筑的异曲同工之妙

Exploring the Divergent Forces of
SWISS ARCHITECTURE

安娜·鲁斯等 | 编

安雪花 于风军 杜丹 孙探春 王京 徐雨晨 焦明 伍晴天 | 译

大连理工大学出版社

04 探索瑞士建筑的异曲同工之妙

004 *探索瑞士建筑的异曲同工之妙*_Anna Roos

010 Caviano的新混凝土住宅_Wespi de Meuron Romeo Architects

030 JS谷仓_Alp'Architecture

040 Refugi Lieptgas建筑_Nickisch Walder

054 Domat/Ems的Tegia da vaut建筑_Gion A. Caminada

062 巴尔斯塔尔某住宅_Pascal Flammer

076 Rotsee终点站_Andreas Fuhrimann Gabrielle Hächler Architekten

086 瑞士鸟类研究所游客中心_mlzd

100 苏黎世瑞士国家博物馆扩建_Christ & Gantenbein

114 新型社区图书馆

114 *规划新式图书馆*_Tom Van Malderen

120 Minna no Mori岐阜媒体中心_Toyo Ito & Associates, Architects

136 Saint-Just Saint-Rambert中央图书馆_Gautier + Conquet

148 Constitución公共图书馆_Sebastián Irarrázaval Arquitectos

162 海岸图书馆_Vector Architects

182 建筑师索引

Exploring the Divergent Forces of
SWISS ARCHITECTURE

004 *Exploring the Divergent Forces of Swiss Architecture _ Anna Roos*

010 New Concrete House in Caviano _ Wespi de Meuron Romeo Architects

030 JS' Barn _ Alp'Architecture

040 Refugi Lieptgas _ Nickisch Walder

054 Tegia da vaut in Domat/Ems _ Gion A. Caminada

062 House in Balsthal _ Pascal Flammer

076 Rotsee Finish Tower _ Andreas Fuhrimann Gabrielle Hächler Architekten

086 Swiss Ornithological Institute Visitor Center _ mlzd

100 Swiss National Museum Extension Zürich _ Christ & Gantenbein

New Community Library

114 *Planning the New Library _ Tom Van Malderen*

120 Minna no Mori Gifu Media Cosmos _ Toyo Ito & Associates, Architects

136 Central Library in Saint-Just Saint-Rambert _ Gautier + Conquet

148 Constitución Public Library _ Sebastián Irarrázaval Arquitectos

162 Seashore Library _ Vector Architects

182 Index

探索瑞士建筑的异曲同工之妙
Exploring the Divergent Forces of
SWISS AR

瑞士建筑一直享有传承高品质和完整性的美誉。这种美誉是通过数个世纪历代工匠和工程师们的努力获得的。瑞士建筑注重细节，对材料和施工有深入的了解，并且能够在特定的历史和文化背景下进行技术创新。

本文将对瑞士建筑的文化以及促进其设计优良、建造一流的建筑的各种力量进行探索。现代瑞士建筑师如何吸收丰富的建筑遗产，同时从科技所提供的最新工具中获得灵感呢？在现代背景下，如何使用和重新诠释古代建筑技术？自然资源的利用对施工所起的作用有多大？对这些问题的探索揭示了各种因素间复杂的相互作用，正是这些因素促使瑞士一批影响深远的建筑物的出现，其影响力超出瑞士这个小小的内陆国，辐射海内外。

Swiss architecture has a reputation for conveying quality and integrity. This status has been earned over centuries through generations of craftsmanship and engineering. Architecture in Switzerland displays an attention to detail, an intimate knowledge of materials and construction, and an ability to innovate within a specific historical and cultural context.

The culture of architecture in Switzerland and the divergent forces that propel the production of well-designed and well-executed buildings will be assessed in this essay. How do the current generation of Swiss architects draw on their rich architectural heritage, whilst at the same time gain inspiration from the latest tools that technology has to offer? How are age-old building techniques used and reinterpreted in a modern context? To what extent does the use of natural resources play a role in construction? The exploration of these questions reveals an intricate interplay of forces that have resulted in a collection of powerful buildings whose influence radiates beyond the confines of the small, landlocked country.

Caviano的新混凝土住宅_New Concrete House in Caviano/Wespi de Meuron Romeo Architects
JS谷仓_JS' Barn/Alp'Architecture
Refugi Lieptgas建筑_Refugi Lieptgas/Nickisch Walder
Domat/Ems的Tegia da vaut建筑_Tegia da vaut in Domat/Ems/Gion A. Caminada
巴尔斯塔尔某住宅_House in Balsthal/Pascal Flammer
Rotsee终点站_Rotsee Finish Tower/Andreas Fuhrimann Gabrielle Hächler Architekten
瑞士鸟类研究所游客中心_Swiss Ornithological Institute Visitor Center/mlzd
苏黎世瑞士国家博物馆扩建_Swiss National Museum Extension Zürich/Christ & Gantenbein

探索瑞士建筑的异曲同工之妙_Exploring the Divergent Forces of Swiss Architecture/Anna Roos

CHITECTURE

瑞士已经成为最前沿工程，高品质设计和精湛工艺的代名词。全世界的建筑爱好者都前往瑞士去欣赏位于苏黎世和巴塞尔的那些具有代表性的建筑，去位于阿尔卑斯山的偏远山谷寻找灵感。在巴塞尔，游客可以游览城市建筑。在这里，你可以接连看到刊登在建筑杂志上的建筑。大量的天才得到了世界的认可，美名远播到世界各地。设计和建筑已经成为瑞士最畅销的出口产品之一。

像彼得·卒姆托和赫尔佐格&德梅隆这样的瑞士建筑师已经收到远至中国、俄罗斯和美国的邀约，由他们来设计奥运场馆和博物馆。他们的设计激发了全世界建筑师的灵感。这并不是说瑞士设计的每一座建筑都很出色；当然，瑞士也有相当多的陈腐平庸的普通建筑。但是，即使是这些不起眼的建筑，其细节往往都是经过非常精心的设计和建造。

在瑞士，众多精美的建筑并不是凭空而建的。它们是几代人努力的结果，众多因素交织在一起才形成了高品质的建筑。建筑艺术已经扎根于瑞士很多个世纪了。这个国家具有避免冲突的悠久历史，这使其建筑物持续而长久，历经数百年都没受到战争的破坏，保存完整。瑞士有四种官方语言——瑞士德语（每个谷地都有自己的方言）、法语、意大利语和罗曼什语——和四个不同的文化群体。瑞士人能够流利地说三到四种语言很正常。其种族的多样性也在建筑遗迹的多样性中得以体现。每个地理区域都有其独特的地方建筑。每一座本土建筑都集合了周围地区丰富的自然资源的不同特点。历史上，每个地区都有其独特的住宅：无处不在的高山木屋，在恩加丁用石头所建的意大利风格的别墅，或者是提契诺州所使用的粗糙的花岗岩和石板。这一丰富的建筑传统被看作是一种珍贵的资源。

Switzerland has become synonymous with cutting edge engineering, high-quality design and precise craftsmanship. Worldwide architecture devotees travel to Switzerland to admire exemplary buildings in Zürich and Basel and to visit remote valleys in the Alps for inspiration. Basel offers architecture tours and in the city you see one building after another that has been featured in architecture journals. The abundance of talent has been recognized and disseminated outside the borders of the small country. Design and architecture have become one of Switzerland's most successful exports. Swiss architects like Peter Zumthor and Herzog & de Meuron have received commissions for Olympic stadia and museums from countries as far afield as China, Russia and the United States and their work inspires architects the world over. This is not to say that every building designed in Switzerland is remarkable; of course there is a considerable volume of banal, commonplace buildings. But even these unexceptional buildings tend to be carefully detailed and well built.

The wealth of fine buildings in Switzerland has not arisen in a vacuum, but has evolved over generations thanks to a complex web of factors woven together to support the production of high-quality architecture. The art of architecture has been cultivated for many centuries in Switzerland. It is a country with a long history of avoiding conflict, allowing continuity and longevity of buildings remaining intact for hundreds of years without being destroyed in wars. With four national languages – Swiss German (each valley has their own dialect), French, Italian and Romansh – and four distinct cultural groups, it is common for Swiss people to speak three or four languages fluently. The diversity of its people is reflected in the variety of its architectural heritage. Each geographic area has its own distinct vernacular architecture, each

很多小规模建筑事务所推动了瑞士建筑的多样性。为"Baukultur"提供发展动力的是完善的教育体制和开放竞争的传统。只有通过公开竞争,才能获得公共建筑项目的设计委托,这让一些新的、名不见经传的建筑设计人才脱颖而出,同时也给有资历的建筑师增加压力,推动他们不断进步。

虽然瑞士建筑作品的多样性很难分类,但是从当代瑞士建筑师的设计中都可以发现惯用的设计理念和一致的设计方法。无论是在国内还是国外,瑞士建筑师设计和建造的大量建筑中无不体现了其对细节的注重,对材料和结构的深入了解以及所具有的创新能力。这些特质也同样体现在瑞士的其他产业方面,如制表业、制药业和纺织业。在这个内聚的方法中,也体现了影响力和方法论的二元性,两者相互排斥,相互制衡。

在瑞士当代建筑这一大范围上,人们可以发现其中颇为有趣的波动。一方面,有像吉昂·A·卡米纳达这样的建筑师,他们认真研究当地具体条件下的富有历史的建筑先例,其设计的作品让人感官非常愉悦、精雕细琢;另一方面,又有像Valerio Olgiati和赫尔佐格&德梅隆这样激进的建筑师,他们的设计放眼全球,其作品极为抽象。像卡米纳达这样的建筑师坚信,建筑项目要紧密贴近所在社区,所用的建筑材料和建筑方法也要本土化。他们支持拥护传统技艺和当地手艺。卡米纳达所设计的优雅的Tegia da Vaut项目,掩映在树林之间,是体现上述设计理念的代表作品。像巴涅地区的谷仓改造和作为本期C3专题设计案例的Tegia da Vaut森林木屋这样的建筑,都集中体现了瑞士建筑师对本土建筑和地方手工艺的尊重。Alp'建筑事务所让他们的木匠将有整整二百年历史的旧木谷仓拆卸下来,重新设计,把这个曾经用作牛

incorporating a distinct palette of natural materials abundant in the vicinity. Historically, each region has its own unique response to dwellings: the ubiquitous alpine chalets in timber, Italianesque villas in Engadin in stone, or rough granite and slate in Ticino. This rich architectural tradition is regarded as a precious resource.

Switzerland's architectural variety is driven by the great number of small-scale architectural practices. The momentum fuelling this "Baukultur" is the excellent education system and the tradition of open competitions. Commissions for public projects are awarded via an open competitive process, allowing the unknown, new talent to rise up the ranks, whilst also keeping the pressure on established architects to advance constantly.

Although the diversity of work produced in Switzerland defies easy categorisation, there are recurrent ideas and consistent approach that runs through the designs produced by contemporary Swiss architects. The attention to detail, intimate knowledge of materials and construction, and ability to innovate – attributes also seen in other Swiss industries like watchmaking, pharmaceuticals and textiles – is evident in the large number of buildings designed and constructed by Swiss architects locally and internationally. Within this cohesive approach there is also a duality of influences and methodologies that move in opposing directions, counterbalancing one another.

The broad palette of contemporary architecture in Switzerland displays an intriguing oscillation between, on one hand the sensual, highly crafted work by architects like Gion. A Caminada, who scrutinise historical precedents in a specific local setting and, on the other hand, the intensely abstract work by radical architects like Valerio Olgiati and Herzog & de Meuron who work in a global setting. Architects like Caminada strongly believe in building projects in close proximity to their own community with materials and skills drawn from the locality. They champion traditional techniques and

Caviano的新混凝土住宅,瑞士
New Concrete House in Caviano, Switzerland

棚和储存干草的旧谷仓重新建成了一个可爱的新家。此建筑一丝不苟地采用了被称为"Strickbau"的古老的木结构建筑技艺,坚固的木柱子横向紧密地结合在一起。在巴尔斯塔尔,由Pascal Flammer设计的曝光度颇高的木屋,其原型也可追溯到这一古代典型的"ur-cabin"建筑形式。细看一下,无论是Flammer设计的小木屋还是Alp'建筑事务所对木质谷仓的改造,都不只是对年代久远的瑞士牧人小屋稍作处理修改的模仿复制,两者都例证了许多瑞士建筑师能够把当代地方因素融入到古老的类型学中。建筑学教授Miroslav Šik首创"altneu"一词来命名这一建筑类型,意为新旧类型学。

瑞士也有高度抽象、充满活力的建筑,与这种"慢建筑"形成鲜明对比。新近开放的由Christ & Gantenbein设计的苏黎世瑞士国家博物的扩建部分就是其中之一。新增加的部分弯弯曲曲,就像复杂的手工折纸,将建于19世纪的博物馆的翼楼与另一座翼楼连为一体。虽然增建部分与这座建于1898年的博物馆的历史建筑风格形成鲜明对照,但是新建筑也模仿了老建筑那带有尖尖的顶点和波谷的复杂的屋顶景观。新旧之间有着千丝万缕的联系,就像在双人舞中,两者相互依赖。新旧并置的设计极其大胆,勇气可嘉,也使两者相得益彰。Christ & Gantenbein的这一设计也充分体现了混凝土在瑞士建筑设计和建造方面的重要性,过去如此,现在也是如此。混凝土本身固有的可塑性是值得肯定的,几乎可以让建筑物呈现各种形式。博物馆扩建部分的混凝土表面从里到外都没有做任何的装饰。唯一的"装饰"就是精美的、垂直安装的日光灯和镶嵌在宽敞气派的楼梯两侧的、光滑的、烫金的、锃亮的铜扶手。不像很多高端大气的"标志性"建筑,这个酷酷的无任何装饰的空间能够让展品占据应有的突出位置,成为焦点,

local crafts. Caminada's elegant Tegia da Vaut nestled in the woods typifies these ideals. Buildings like the barn renovation in Bagnes and the forest hut, Tegia da Vaut featured in this issue of C3 epitomize the deference that Swiss architects pay to their vernacular architecture and local craftsmanship. Alp'Architecture had the entire two-hundred-year-old timber barn – former cow shelter and hay storage – dismantled and coded by their carpenter to be lovingly reused in the reconstruction of the old barn into a new home. The ancient timber construction technique "Strickbau" using solid timber posts knitted together horizontally has been meticulously adopted. Pascal Flammer's highly publicised timber House in Balsthal owes its archetypical form to the ancient "ur-cabin". On second glance, though neither the Flammer nor Alp' chalet is in any way a mere rehashed replica of the clichéd Swiss chalet, but are both examples of the ability of many Swiss architects to interpret their contemporary vernacular elements into the old typology. Architecture professor, Miroslav Šik coined this typology of architecture "altneu" or old-new.

In stark contrast to this kind of "slow architecture", there are highly abstract, dynamic buildings like Christ&Gantenbein's newly opened extension to the Swiss National Museum Zürich. The new addition zigzags its way like a complex piece of folded origami, linking one wing of the 19th century museum to another. Although it sharply contrasts with the historic style of the 1898 building, the new wing also mimics the complex roofscape of sharp peaks and troughs of the old building. The old and new are inextricably linked, as if in a dance duet, dependant upon each other. The juxtaposition is an extremely bold and daring gesture that heightens both old and new. Christ&Gantenbein's project is also a fine example of the importance of concrete in designing and constructing buildings in Switzerland both in the past and present. The innate plasticity of concrete is embraced, allowing buildings to be sculpted in almost any form. Both in and out, the concrete surfaces have been left raw, the only "decoration" being the

瑞士鸟类研究所游客中心，森帕赫
Swiss Ornithological Institute Visitor Center in Sempach

而不会让任何花哨的建筑方面的元素分散参观者的注意力。

习俗和传统在瑞士建筑中发挥了核心作用，与之相对应的原创性和创新性也是如此。很多建筑看起来非常奇特，又古色古香。Wespi de Meuron设计的位于Caviano的新混凝土住宅就是其中之一。新混凝土住宅建于马焦雷湖旁的小山坡上，顺着地形拾级而下。从下面往上看，粗糙的、若隐若现的外立面带有分散的开口设计，看起来就像是一个天然的悬崖，在岁月里饱经风霜的洗礼。Nickisch Walder建筑事务所设计的Refugi Lieptgas建筑既展示出许多瑞士建筑师对建筑遗产的敏感，又展示出他们的创新能力。通过使用从原来旧的小木屋中拆下来的木头托梁作为新建筑的模架，建筑师创造出了精美的圆齿状边缘，以此铭记原来小木屋的木质结构。原来木屋的轮廓在新建筑上留下了不可磨灭的痕迹，就像古化石的印记。

由Andreas Fuhrimann Gabrielle Hächler建筑师事务所设计的Rotsee终点站坐落在卢塞恩附近的湖面上，集功能性和雕塑般的美感于一身，也是本期重点推出的设计案例。预制木结构元素一个一个堆砌起来，就像用小孩玩具箱里的积木搭建起来的一样。使用时，大幅的百叶窗慢慢打开并弹出；在冬季不用的时候，这个塔就关着，处于休眠状态，直到举行一年一度的夏季赛艇大赛时才打开。这座小建筑设计新颖，理念创新。

在瑞士，人们在设计上都会注重其功能性和可持续性，"Minergie"可持续建筑标准得到建筑师和客户们的广泛认可和支持（银行甚至给予符合Minergie建筑标准的项目抵押贷款方面的折扣）。从一方面来说，瑞士建筑师既考虑使用尖端的技术和精湛的工程设计，另一方面，他们也使用低技术含量的生态方法和材料。瑞士鸟类研

delicate vertical fluorescent light fittings and the smooth burnt-gold of the burnished brass hand rails that finely frame the monumental stairs. Unlike so many brash "signature" buildings, the cool, naked spaces allow the exhibits the prominence and focus they deserve without any gimmicky architectural distractions.

Convention and tradition play a central role in Swiss architecture, as do the opposite: originality and innovation. Many buildings, like Wespi de Meuron's New Concrete House in Caviano cascading down the hill above Lago Maggiore have an eerie, archaic quality. From below, its rough, looming facades with its scattered openings read like a natural cliff-face eroded by the elements over time. Nickisch Walder's Refugi Lieptgas displays both the sensitivity that many Swiss architects have for their built heritage and their ability to innovate. By reusing the original timber joists from the old chalet as formwork for the new building, the architects created a delicate scalloped edge that remembers the timbers of the old chalet. The old timber contours have left an indelible stamp on the new, like the imprints of ancient fossil.

Andreas Fuhrimann Gabrielle Hächler's Rotsee Finish Tower on a lake near Lucerne – also featured in this issue – is both functional and sculptural at the same time. Built from prefabricated timber elements stacked one above the other, it is rather like the wooden blocks from a child's toy box. When the building is in use, its large-format shutters are slid open and popped up, during the winter when not in use, the tower closes up remaining dormant until the annual summer rowing regatta. This small building is both original in its design and innovative in its conception.

Functionality and sustainability are esteemed in design in Switzerland, with "Minergie" standards for green architecture being widely recognized and supported by architects and their clients. (Banks even give mortgage discounts to Minergie buildings). Swiss architects embrace both cutting edge technology and sophisticated engineering on one hand, and

Rotsee终点站，Lucerne，瑞士
Rotsee Finish Tower in Lucerne, Switzerland

究所的mlzd新游客中心的外壳由带有有机纹理的夯土实心墙构成。其设计如同赫尔佐格·德梅隆建筑事务所在2014年设计的利口乐仓库——欧洲最大的夯土结构。

同这种务实、低调的做法相对应的是智慧思考这一深层流，许多瑞士建筑师的作品都受到深深的影响。像Lars Müller和Birkhäuser这样的出版社出版了许多尖端的研究建筑和城市规划的理论和实践方面的书籍。这些书都是智慧思考的成果。与这种几乎坚忍、务实的设计态度完全相反，一些瑞士建筑师渴望把纯艺术的成分添加到建筑设计中。许多备受瞩目的瑞士建筑师往往都愿意与优秀的艺术家合作，就是为了能够用与众不同的纹理、色彩和精美的图像来丰富他们的建筑设计。实际上，他们把建筑的外立面看成是巨大的画布。

无论是赫尔佐格·德梅隆建筑事务所早期设计的、建筑外覆层是带有浮雕图案的混凝土和玻璃的利口乐仓库项目，还是由Diener+Diener建筑师事务所设计的诺华园的熠熠发亮的玻璃外墙，都是精美艺术与建筑交叉结合的典型例子。许多瑞士建筑师深深置身于其建筑遗产之中，然而也决不畏惧尝试运用最新技术和新颖独创的建筑形式。受到众多因素影响，总的来说，瑞士建筑很好地保持了过去知识和当下创新之间、本土建筑方法和激进设计之间、永恒性和时代性之间、理论和实践之间、理性和直觉之间以及谦逊和大胆之间的平衡。上述所描述的影响瑞士建筑不同因素之间的错综复杂的相互作用产生了一批有影响力的建筑，其影响力远远超出了这个小小的内陆国家的国界，辐射海外。

low-tech ecological methods and materials on the other. The outer shell of mlzd's visitor's center at the Swiss Ornithological Institute Visitor Center is comprised of organically textured, solid rammed earth walls, as is Herzog & de Meuron's 2014 Ricola storage building – the largest rammed earth structure in Europe.

The counterbalance to this pragmatic, low-key approach is the deep current of intellectual thought that courses through the work of many Swiss architects. Publishers like Lars Müller and Birkhäuser publish sophisticated volumes delving into theory and practice of architecture and urban planning. These books are the fruits of this discourse. Also running contrary to the almost stoic, practical stance to design is the desire of some Swiss architects to add pure artistic flair to their buildings. A number of high-profile Swiss architects have a penchant for collaborating with fine artists, enriching their buildings with unusual textures, daubs of colour and delicate images, basically treating their facades like vast canvases.

Herzog de Meuron's earlier Ricola project with its embossed concrete and glasswork cladding, or the shimmering glazed facade of Diener + Diener's Novartis Campus are pertinent examples of this cross pollination of fine art and architecture. Many Swiss architects are deeply rooted in their heritage, yet unafraid to experiment with the latest technology and original forms. Within the multitude of influences, a broad current of Swiss architecture demonstrates a fine equilibrium between past knowledge and current innovation, vernacular methodology and radical design, timelessness and contemporaneity, theory and practice, rationality and intuition, as well as modesty and boldness. The intricate interplay between the divergent forces described above has resulted in a collection of powerful buildings whose influence radiates far beyond the confines of the small, landlocked country. *Anna Roos*

探索瑞士建筑的异曲同工之妙　Exploring the Divergent Forces of Swiss Architecture

Caviano的新混凝土住宅
Wespi de Meuron Romeo Architects

该建筑是一个三口之家的住所，紧邻马焦雷湖沿岸的Caviano区建筑事务所。

1981年，该建筑事务所在这个仅有128m²的地块上建造。从要有足够的建筑密度的需求来说，如果不破坏现有的环境，需要在这处地块的其余地方开拓新的居住空间。反之，考虑到现有的建筑，要想有足够的建筑密度，就应该改善目前的外部空间状况。

每当开发剩余的地块时，建筑法常常决定了建筑的外部形式。房屋与道路的最小距离、与森林的最小距离、与建筑事务所的最小距离、西南邻居边界的建房局限，这些都决定了这座表面积为79m²的建筑不规则的五角形外观。

这个不规则的外形之内有清晰的、面积为48m²的长方形表面，形成独立的内部空间。

多边形的建筑外形和场地陡峭的地形使整座建筑看起来就像森林中央的一块古老的石块。粗糙的混凝土表面经过风雨的冲洗还会变暗，更加凸显了这种感觉。

从靠近山边的街上看过去，这栋建筑只是一个封闭的、只有一层的建筑体量。街道一侧唯一的洞口是通向入口院子的由粗钢建成的钢门。

3m宽的前院的路面由天然石头铺成，种有两棵棕榈树，院子将房屋和街道连接起来，并且提升了其空间性。而从山谷一侧望去，该住宅呈现为一座狭窄的三层塔楼。

房屋有三层：顶层位于街道层，房屋的入口、主客厅和带开放厨房的餐厅都位于这一层。顶层的其中两侧是完全封闭的，另外两侧完全由玻璃覆盖，面向庭院。

阳光从楼梯上方的天窗照进下面一层。这一层有两间卧室、浴室和通向地窖的楼梯。每个卧室都有自己的户外凉廊，地窖内设一个设备间和一个工作间。

两个庭院都长满紫藤，使客厅像位于花园中一样，居住者也能以独特的方式体验到天气和光线所带来的千变万化的气氛。

山边一侧的入口庭院保护房屋免于遭到来自街上的窥探，同时使房屋充满光线。

靠近湖边一侧的内部庭院可以使人们通过一个巨大的屋顶洞口来饱览湖光山色；同时其封闭的墙面还可以将阳光反射进庭院内。

New Concrete House in Caviano

The house, designed as a residence for a family of three persons, was built in the immediate proximity of the architecture office in Caviano on the Lake Maggiore.

In terms of adequate architectural densification new living space had to be created on the remaining area of just 128m², on the same plot as the architecture office was built in 1981, without damaging the existing conditions. On the contrary, an enrichment of the outer spatial situation should be generated with reasonable densification in context with the existing building.

The building laws determined the outer form of the building, what often happens when leftover plots are developed. The minimal distance to the road, the minimal distance to the forest, the minimal building distance to the architecture office as well as the right to build on the limit line to the southwest neighbor, create an irregular pentagonal form of 79m² surface in total.

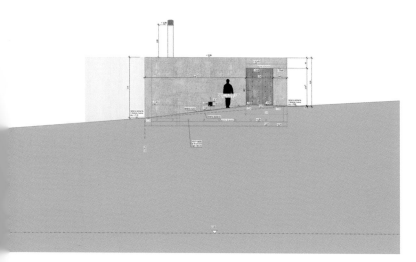

南立面 south elevation

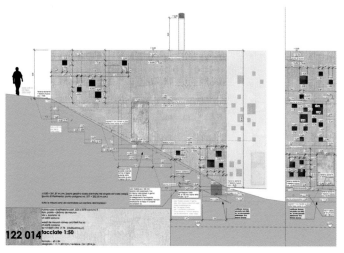

东立面 east elevation

项目名称：New Concrete House in Caviano
地点：6578 Caviano, Gambarogno, Tessin, Schweiz
建筑师：Wespi de Meuron Romeo Architects
承包商：Fam. Jérôme + Paola de Meuron
工程师：de Giorgi & Partners, 6600 Muralto
建筑物理：IFEC Ingegneria SA, 6802 Rivera
用地面积：128m²
表面积：79m²
有效楼层面积：80m²(net, without walls, 2+3 floor)
造价：CHF 750,000
设计时间：2013
施工时间：2014
竣工时间：2015.1
摄影师：©Hannes Henz(courtesy of the architect)

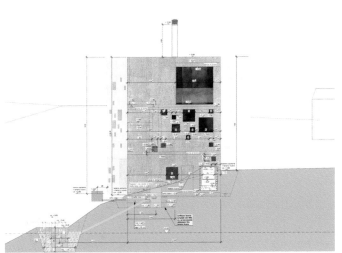

北立面 north elevation

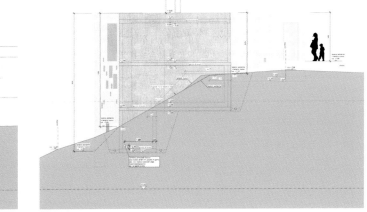

西立面 west elevation

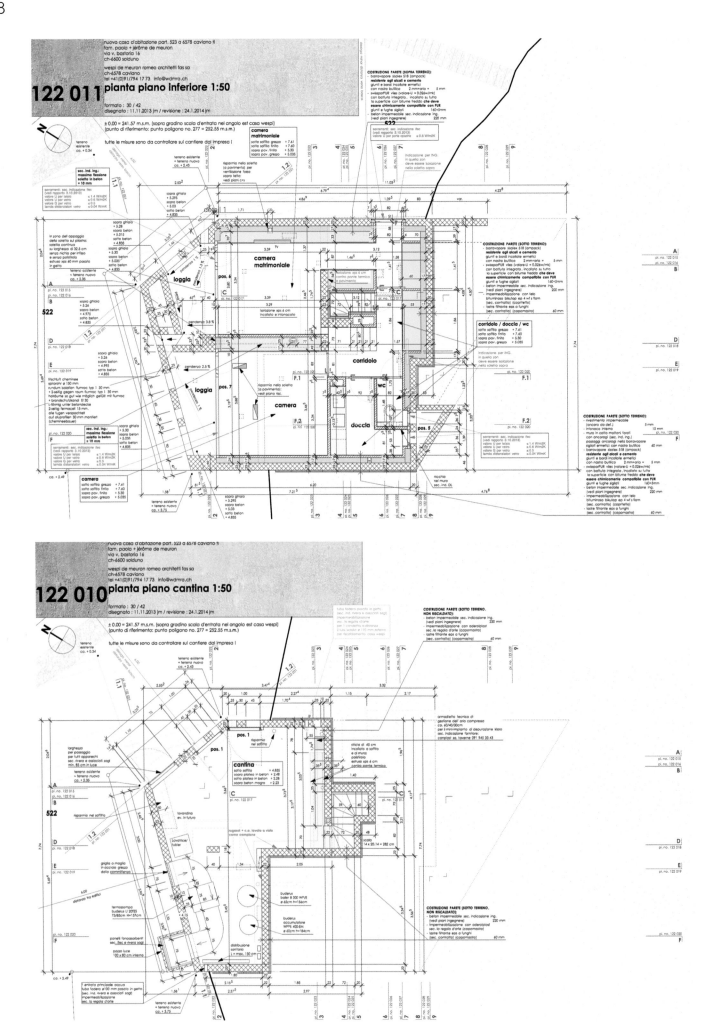

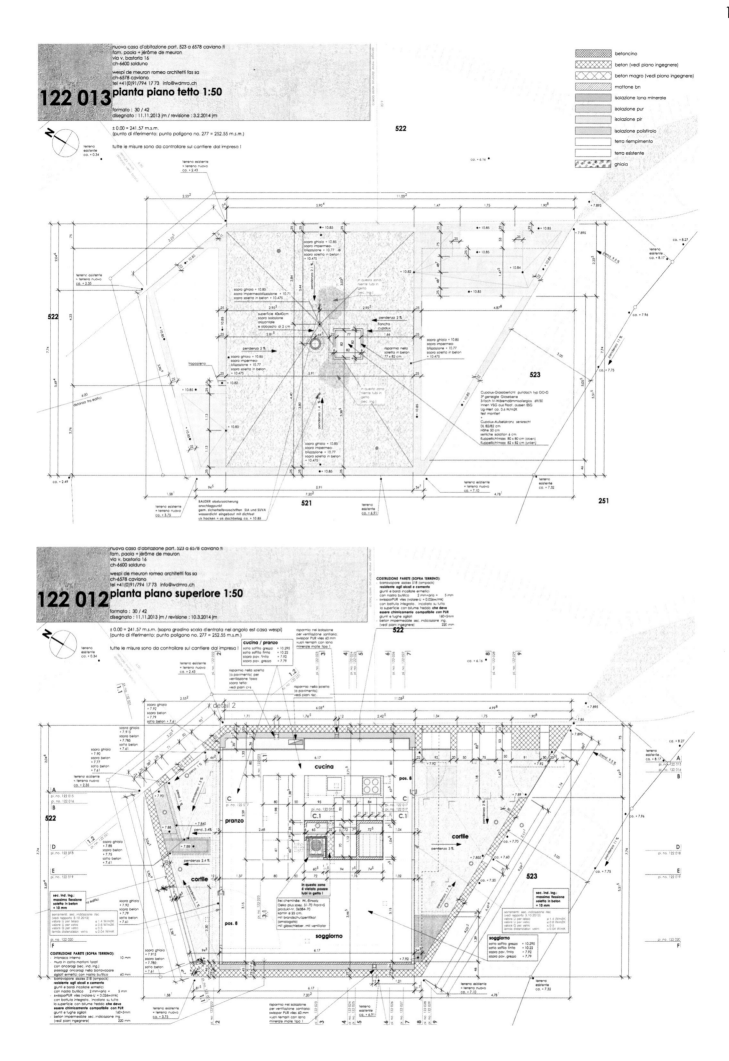

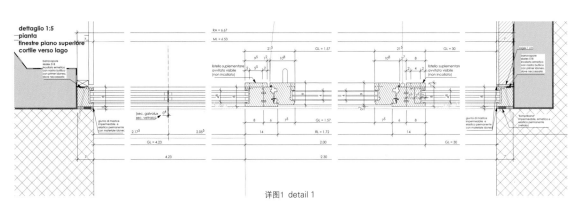

详图1 detail 1

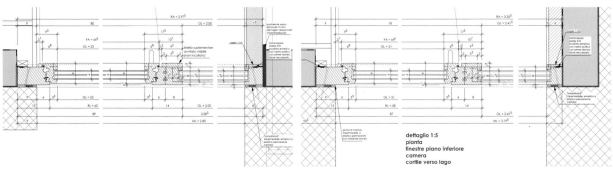

详图2 detail 2

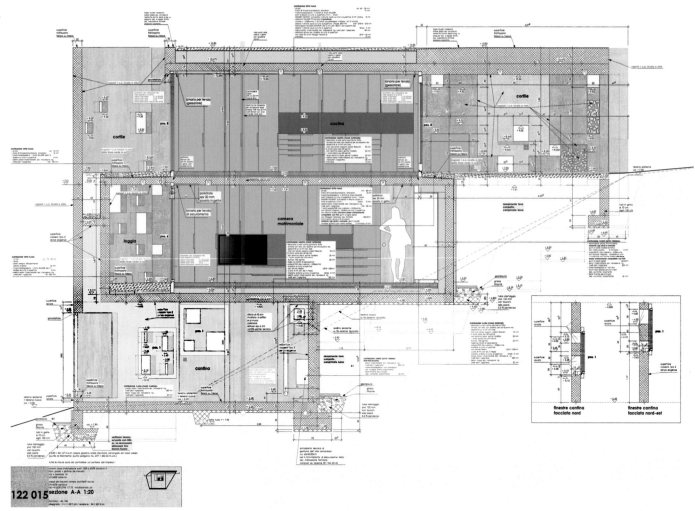

122 015 sezione A-A 1:20

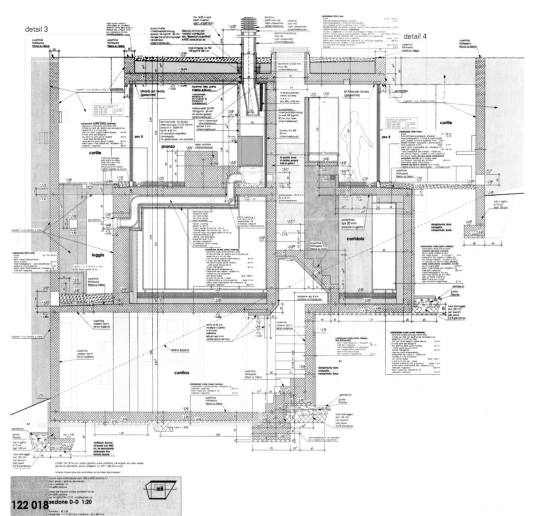

122 018 sezione D-D 1:20

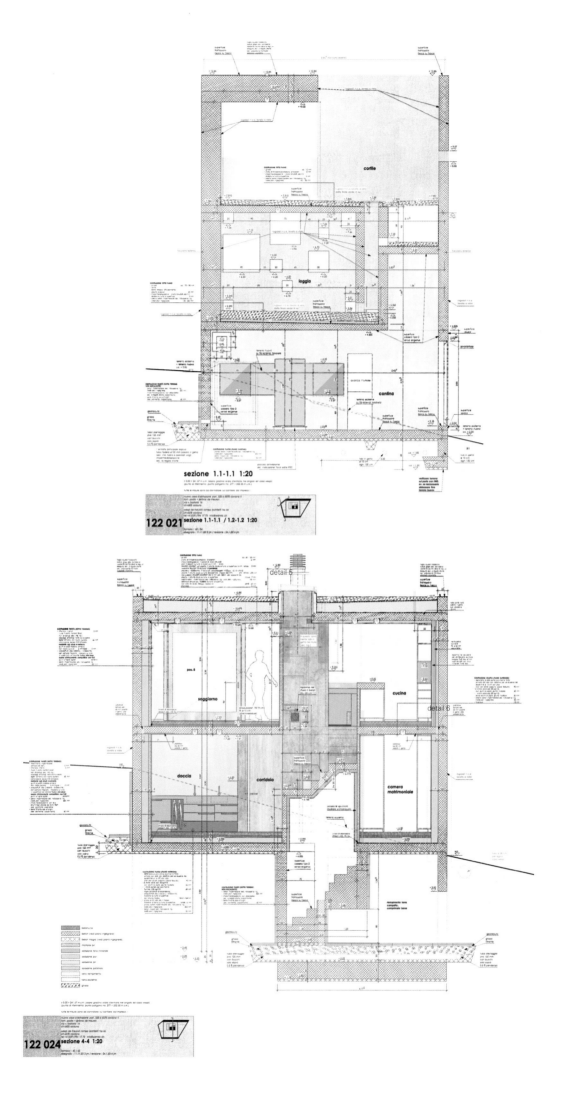

A clear rectangle of 48m² surface, which is the isolated interior, was integrated in this irregular form.

The polygonal exterior shape and the steep topography of the site let the building appear as an archaic stone block in the middle of the forest, reinforced by the rough washed concrete surfaces which will darken as they weather.

To the mountain-sided street the construction presents itself as a closed, simple one-storey volume. The only opening towards the street is the raw steel gate leading to the entrance court.

A 3m wide forecourt with a natural stone pavement and two palms connects the house to the street and upgrades it spatially. To the valley-side, the house appears as a narrow 3-storey tower.

The house is organised on three floors: the top floor on the street level accommodates the entrance, the main living area and dining with the open kitchen, on two sides it's completely closed and on the other two sides it's completely vitrified towards the courtyards.

A skylight above the staircase allows light to penetrate into the lower floor, which accommodates two bedrooms, each with its own outdoor loggia, the bathroom and the stairs to the cellar, where is the technique and work space.

Both courtyards, each with a wisteria, let the living room become a "garden" room and let the inhabitants experience in an unusual intense way of varying atmospheres of the weather and the light.

The entrance courtyard on the mountainside protects the house against view from the street and in the meantime it lets the sunlight in.

The inner courtyard on the lakeside releases the view to the lake and the mountains through a big roofed opening; while it's closed wall surfaces reflect the sunlight to the inside.

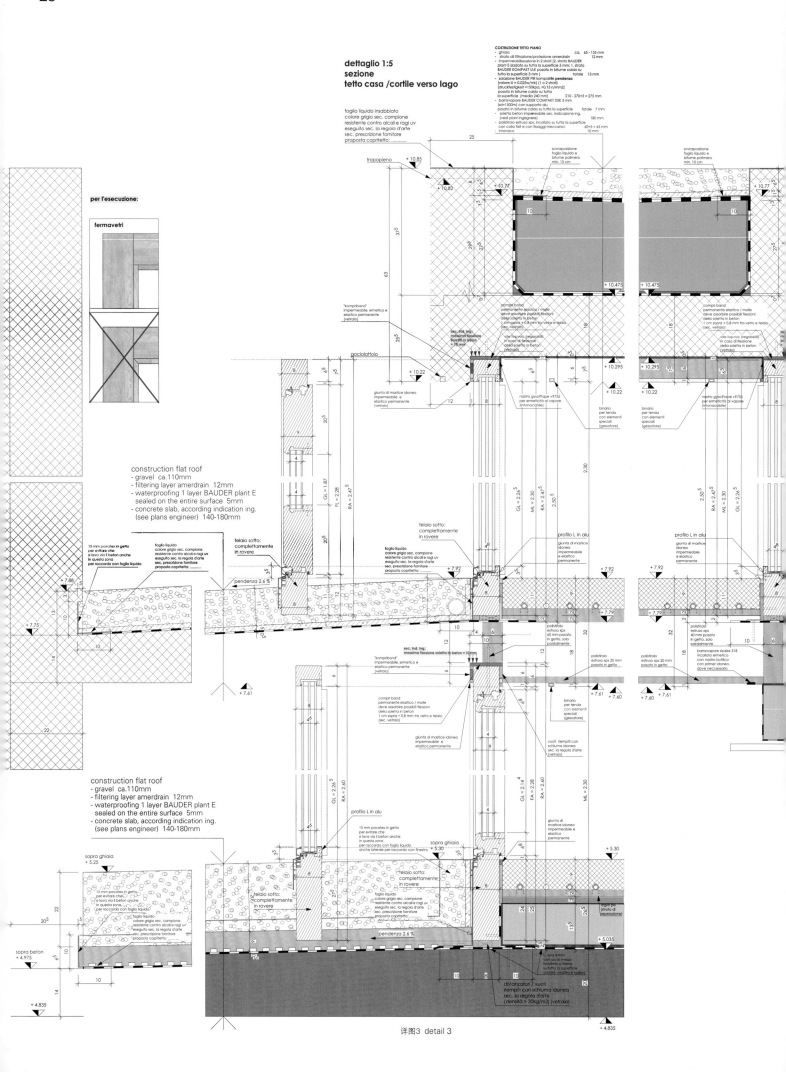

详图3 detail 3

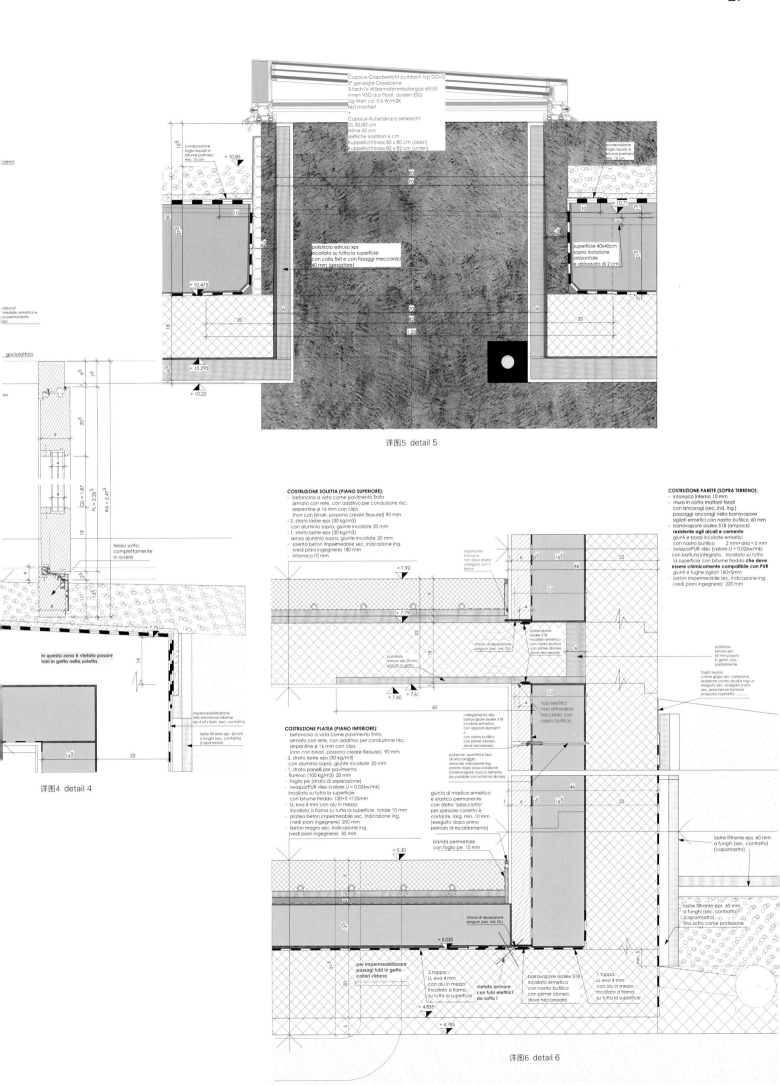

JS谷仓
Alp'Architecture

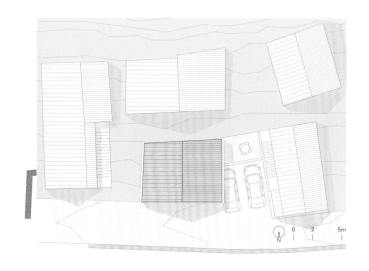

Sarreyer村庄坐落在沃利斯地区巴涅山谷内的一座阳光明媚的山坡上,是处处充满对比的山谷中最典型的村庄之一。

韦尔比亚休闲度假区的活动令人忙乱而兴奋,而设有莫瓦桑大坝的巴涅山谷底部与世隔绝,人迹罕至。而两者之间的中途设有许多谷仓,其建造的时间有的可以追溯至18世纪。Sarreyer村是个适合全年居住的好地方,有很多传统的高山建筑。这个地方的故事与农业有紧密的渊源。

每年都会有许多谷仓被人们翻新改造,其数量表明人们将这些地方作为居住和旅游度假之地的浓厚兴趣,这里既天然又返璞归真。翻新改造这些谷仓为建筑师们提供了将尊重传统与先进的建筑技术和装饰结合起来的好机会。事实上,对这些建筑的适用规定强调要特别尊重现有的建筑体量,当设计新的开窗时要特别注意其外立面与洞口的比率,此外,在保留原有的薄屋顶以及使用尊重场地及其历史的材料方面都有适用规定。

这座谷仓始建于1792年,最初在冬天用作牛棚、晾晒干草和谷物的仓库。200年后,一对年轻的夫妇爱上了这个地方,并买下了这里。经过翻新改造后,这里摇身一变成为一座75㎡的舒适三层小楼。为了建成理想的样子,测量员测量了整座现有建筑,在谷仓拆掉之前,木匠给原建筑中的每块木板标记了序号,以便在新建筑中将它们安放在同样位置。地下室用混凝土重建,而上层的墙体先在工厂车间里加工,然后到现场直接组装。窗户的位置进行了精心设计,且与原建筑不匹配的窗户数目将减到最少。唯一例外的是在北面留了一扇大窗,让阳光洒在这座三层小楼的楼梯上。屋顶也值得特别关注:为了保持原来屋顶单薄的厚度,设计师利用技术将屋顶的框架都隐藏于室内,而外部看不到厚厚的挑檐。

新房主特别注意家具的选择,将原来谷仓里的各种小物件都重新用来装饰新居,或与新居的装饰很好融合在一起。

JS' Barn

Hung on the sunny hillsides of the Valley of Bagnes in Wallis, Sarreyer is one of the most typical villages of this valley full of contrasts.

Halfway between Verbier with its hectic activity and the isolated bottom of the valley with the Mauvoisin dam, numerous barns whose constructions sometimes go back to the 18th century can be found. Sarreyer is a comfortable place to live all year round and hosts many examples of alpine traditional constructions. The story of this site is closely related to agriculture.

项目名称:JS' Barn
地点:Sarreyer, Valley of Bagnes, Wallis, Switzerland
建筑师:Alp'Architecture
项目团队:Sacha Martin, Laurent Berset, Romain Pellissier
工程师:Grégoire Bruchez_LBI Lattion Bruchez ingénieurs SA
用地面积:75㎡
总建筑面积:90㎡
有效楼层面积:65㎡
体积:190㎥
施工时间:2013~2014
摄影师:©Christophe Voisin (courtesy of the architect)

上层 upper floor

底层 lower floor

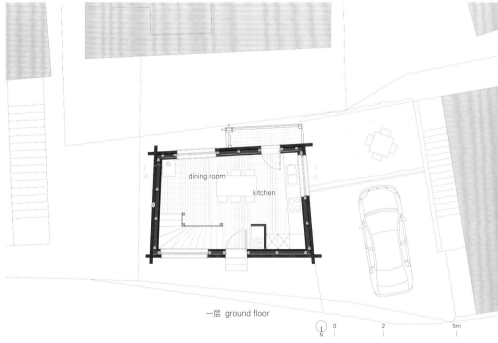

一层 ground floor

The amount of barns renovated each year shows the interest for these areas as residence and vacation places in a context at the same time natural and authentic. These renovations are a unique opportunity for the architects to combine the respect for the traditions with cutting edge construction technology and decoration. In fact, the applicable regulations on these objects impose a drastic respect of the existing volumes and their facades/openings ratios, and a particular care when creating new openings, of the maintenance of a thin roof and the use of materials respectful of the site and its history.

Initially built in 1792, this barn was used to shelter cows during the winter and to dry hay and cereals. Bought 200 years later by a young couple who fell in love with the place, it was renovated and turned into a cozy three-story apartment of 75 square meters. In order to do so, the existing volume was entirely measured by a surveyor and the planks were numbered by a carpenter to be replaced at the exact same spot before the barn was demolished. The basement was then rebuilt with concrete and the upper walls were prepared in a workshop to be reassembled directly on site. Windows were carefully placed and the ones not matching the original openings were reduced to the minimum, with the only exception being a generous window on the north facade, bringing light into the stairs on the three floors. The roof also required a special attention: in order to maintain its low thickness, technical solutions were found to express the framework on the inside without having thick roof overhangs on the outside.

A particular care was given to the furniture and various pieces of the original barn have been reused or integrated into the decoration set up by the new owners.

北立面 north elevation

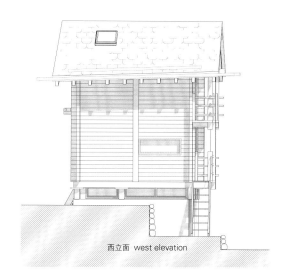
西立面 west elevation

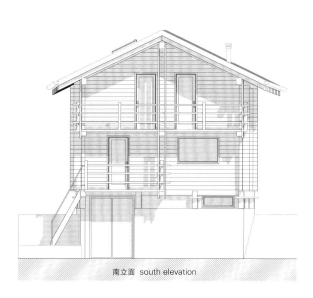
南立面 south elevation

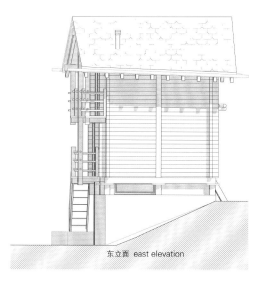
东立面 east elevation

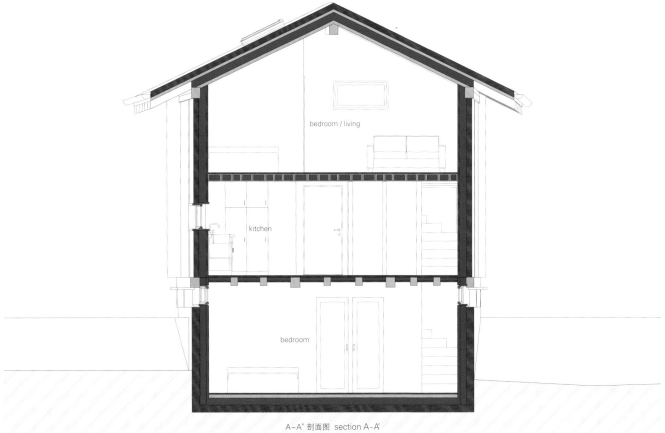

A-A' 剖面图 section A-A'

探索瑞士建筑的异曲同工之妙 Exploring the Divergent Forces of Swiss Architecture

Refugi Lieptgas建筑
Nickisch Walder

Refugi Lieptgas建筑坐落在瑞士阿尔卑斯山区的弗利姆斯。这里是一个旅游景区,如今主要作为一处滑雪胜地。弗利姆斯的特别之处在于它的地形。大约一万年前发生的岩石崩落使山谷内填满了岩石,形成了高原,弗利姆斯就在这高原的边上。标志性的350m高的"Flimserstein"墙位于这个村庄的北部。在南面,一座群山绵延起伏、湖泊点缀于茂密森林之间的乡村一直延伸到瑞士Ruinaulta大峡谷。直到今天,河流还得在石灰石体块中开辟道路,留下了许多美丽的岩石构造。

Refugi Lieptgas建筑的场地以前矗立着一座木结构房屋,用作农民的避难所。小屋前面是一条小路,通向弗利姆斯的森林,其旁边设有一个枯萎的啄木鸟啄空的空心树雕塑。

小屋和马厩都已经废弃,近乎破败。以前住人的小木屋可以经过改造,用新的建筑取代,但这个地方的特色必须保留下来。

新建的小屋就像被遗弃的建筑化石一样,伫立在原地。旧木屋的原木结构仍然作为新建筑的框架,墙体采用加气混凝土浇筑而成,表面很粗糙,自然很快会再次征服这一人工建筑,使其饱经沧桑。

房屋的原有外形决定了楼上的空间。后墙边设有一个壁炉,窗边是餐厅,人们从窗户可以看到林中的一小块绿茵茵的空地,这里还有小厨房和可以加热的混凝土长凳。人们坐在长凳上,透过屋顶圆形的天窗,可以看到一棵大型山毛榉树的树冠,且通过一段弯曲的楼梯,可以到达地下一层。人们可以在此处的寂静中入睡。地下一层的外面是一块巨大的岩石,朦胧的光线只能透过岩石和房屋之间的缝隙进入房间。窗前放着一个浴缸,镶嵌在一个大型加热混凝土基座里。一扇门通向外面。

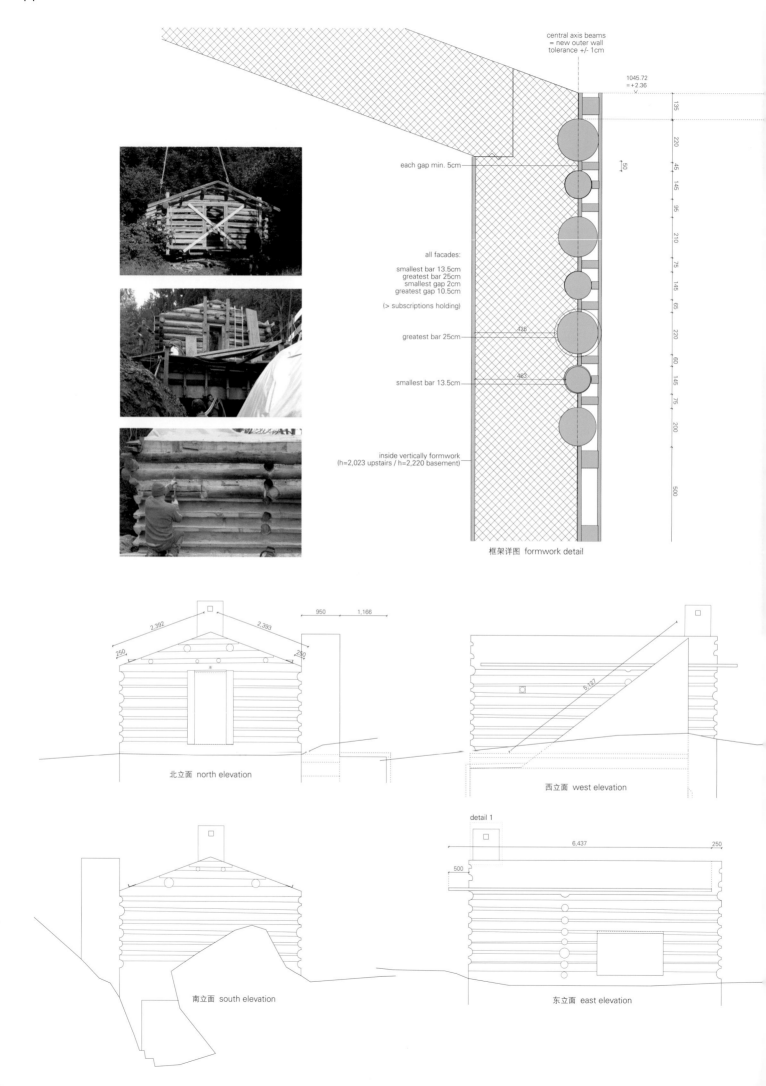

　　小木屋的最初设计目的是供旅游度假客人使用,最多可以住两个人。弗利姆斯的游客大部分时间是在周围游山玩水,小木屋只是一个避难所,一个休息和消遣娱乐的地方。在这里,游客可以享受到高贵、宁静且舒适的生活。不管怎么说,小木屋很小,只能保证最基本的需要,并努力做到保护私密性。所有的元素都根据房屋的大小调整了其大小。小木屋分为两处主要区域,一处是公共区域,用于起居、烹饪、就餐、坐在壁炉前一起聊天;另一处是隐秘的休息场所,用来沐浴和休憩。整座小木屋建筑还包括楼梯、浴室等额外的功能,还有一个没有供暖设施的空间,用来储藏物品和放置为小木屋提供技术支持的机械设备。

Refugi Lieptgas

Refugi Lieptgas is situated in Flims, Swiss Alps. It is a tourist region, today mostly known as the ski and snowboarding resort. The special thing about Flims is the topography. A huge rockslide around 10,000 years ago filled up the valley and formed a plateau, with Flims situated at its border. 350m wall of the iconic "Flimserstein" is located in the north side of the village. In the south, a hilly country densly forested with lakes stretched out to the Ruinaulta, a deep canon. The river had to make his own way through the masses of limerock until today and left some beautiful rock formations.

The refugi Lieptgas stands where in the past a wooden structure served as a shelter to the farming people. In front of the hut, a small path leads into the forest of Flims, next to the sculpture of a hollowed dead woodpecker tree.

Both buildings – cabin and stable – were abandoned, about to decline. The formerly inhabited part of the ensemble could have been converted and replaced by a new construction. But the character of the place had to be preserved.

The new hut now stands as a petrification of the abandoned structure exactly in the same place. The timber log structure of the old cabin was used as the formwork, and the walls were cast in insulating concrete. The surface is rough, and nature will soon reconquer the artefact.

The archetype form of the house defines the space of the upper floor: a fireplace in the backwall, an eating area at a window with view into a clearing of the woods, a small kitchen, and a heated concrete bench from which one looks through a round roof light into the crown of a big beech tree. A sinuous staircase leads to the underground. Here in silence one sleeps. A big rock dominates the space, and crepuscular light falls throughout the gap between rock and house. In front of the window a bathtub is incorporated in a big heated concrete plinth. A door leads outside.

The hut was meant from the beginning for vacation guests, of maximum two people. Guests in Flims do mostly spend their days in the surrounding nature, so the cabin is used as a refuge, a resting and recreation place. A noble and calmly beautiful kind of life should possibly be lived in here. The hut being very small anyway should be reduced to the fundamental, trying to get the quality of a protected intimacy out of the smallness. Everything was resized to this smallness.

The cabin was separated in two essential main spaces – a common one for living, cooking and eating, as well as sitting together at the prominent fireplace, and the other one for retreat and being secluded, for bathing and sleeping. Additional functional rooms like stair, bathroom and an additional non heated space for storage and technical stuff were attached to the cabin volume.

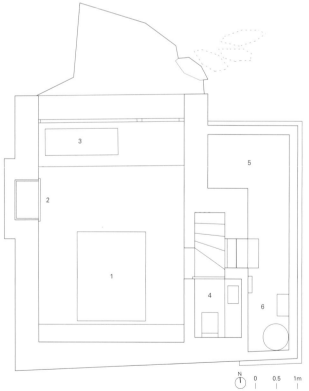

1.床 2.小橱 3.浴缸 4.卫生间 5.储藏室 6.技术空间
1. bed 2. inbuilt cupboard 3. bath tub 4. toilet 5. storage 6. technical space
地下一层 first floor below ground

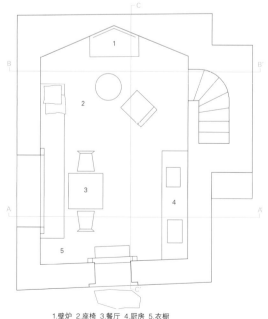

1.壁炉 2.座椅 3.餐厅 4.厨房 5.衣橱
1. fireplace 2. sitting 3. eating 4. kitchen 5. wardrobe
一层 ground floor

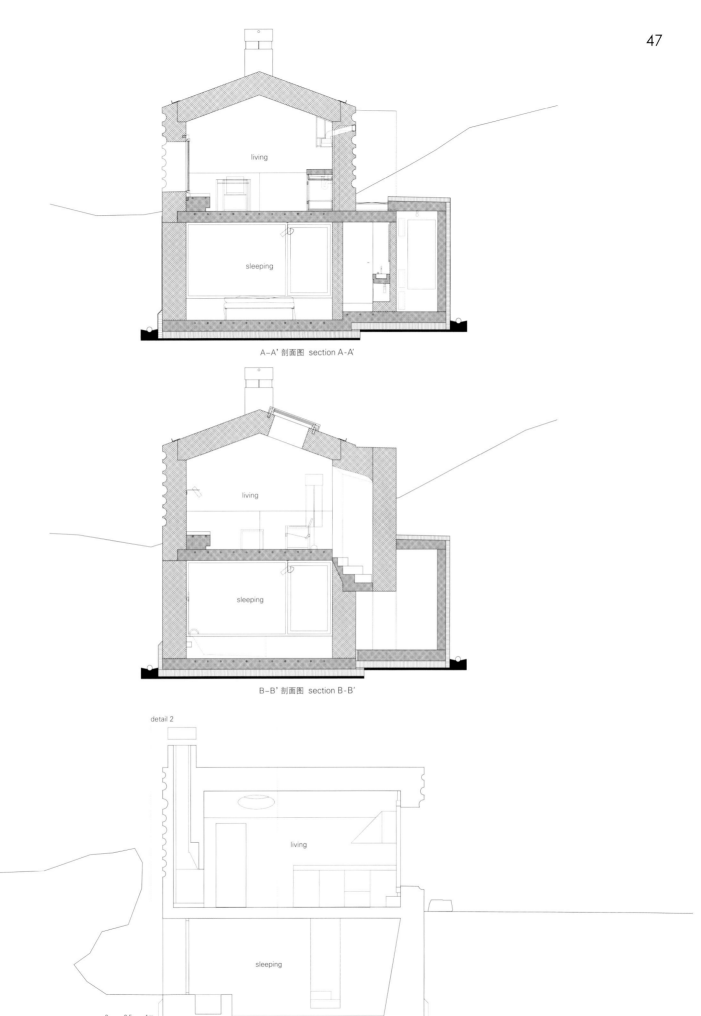

A-A' 剖面图 section A-A'

B-B' 剖面图 section B-B'

C-C' 剖面图 section C-C'

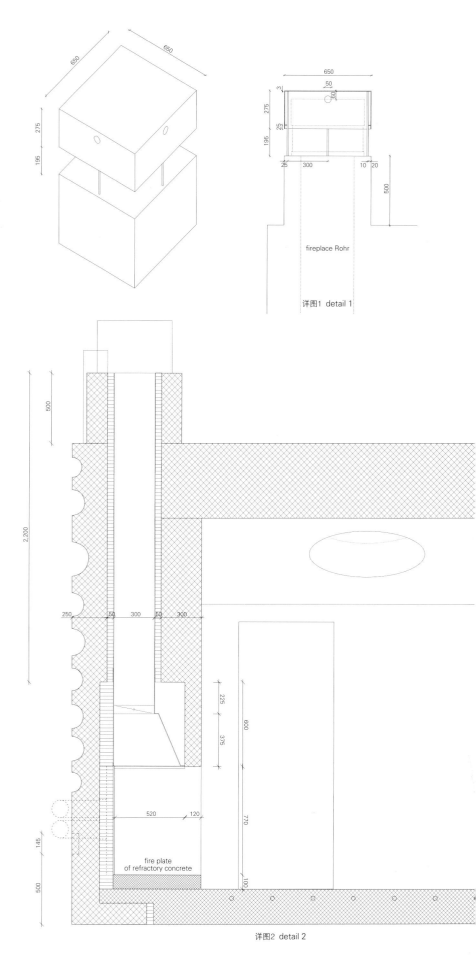

详图1 detail 1

fireplace Rohr

详图2 detail 2

fire plate of refractory concrete

项目名称：Refugi Lieptgas
地点：Flims, Switzerland
建筑师：Nickisch Walder
项目团队：Selina Walder, Georg Nickisch
用地面积：115m² / 总建筑面积：72m²
设计时间：2012 / 竣工时间：2012
摄影师：
©Gaudenz Danuser (courtesy of the architect)
-p.40~41, p.42~43, p.45, p.48~49, p.52
©Ralph Feiner - p.62~63, p.65

Domat/Ems的Tegia da vaut建筑
Gion A. Caminada

探索瑞士建筑的异曲同工之妙 Exploring the Divergent Forces of Swiss Architecture

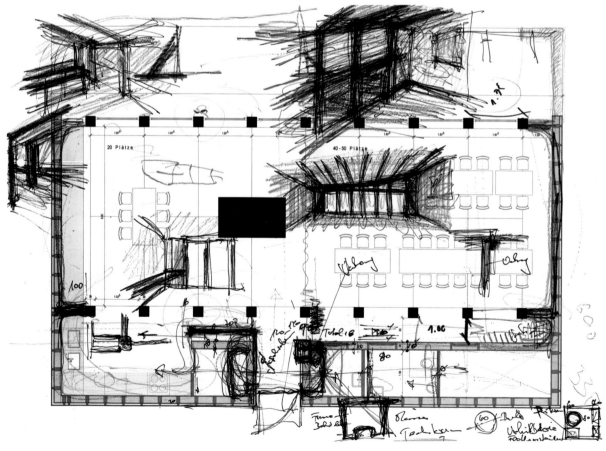

项目名称：Tegia da vaut Domat/Ems / 地点：Domat/Ems, Switzerland
建筑师：Gion A. Caminada / 项目团队：Jan Berni
施工管理：Christiane Bertschi Rageth / 施工工程师：Walter Bieler Ingenieurbüro
客户：Municipality of Domat/Ems / 用途：public space / 造价：1.02 Mio CHF
设计时间：2012 / 竣工时间：2013 / 摄影师：©Ralph Feiner

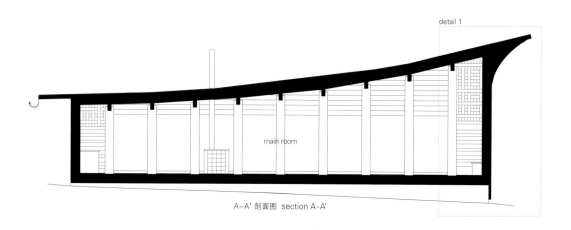

A-A' 剖面图 section A-A'

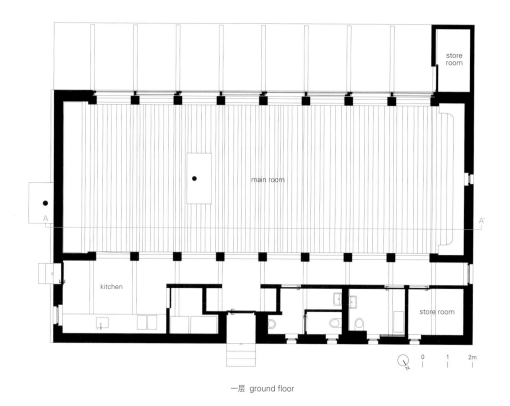

一层 ground floor

　　此项目位于Domat/Ems森林之中，用来观察自然和文化感知现象。森林总是能给人们一些独特的体验，既熟悉又新奇。Tegia da vaut项目坐落在树林中，外形展现为一种乐器的形状，使我们的感官与自然和谐一致。为实现Tegia da vaut的设计理念，建筑师首先从附近收集新建筑的材料，就地取材。这是因为人们相信，一个与众不同的、拥有特定地点的建筑文化的中心应当使用当地的材料。多样性只能通过设定界线来展现并得以保持。

　　在设计Tegia da vaut项目时，建筑师的目的是将一棵树物尽其用。这就意味着树的每一个部分都是有用的。不同的木材将分别发挥加固、支撑、交错连接和分离的作用。在一些地方，木材看起来是一层层的，很沉重；而在其他地方，木材相互交织在一起，又变得很轻盈。其目的是营造完整的空间氛围。用羊毛做保温处理，用皂石烤炉带来浓浓暖意，这些做法都大大影响了人们对这栋建筑独特的整体感觉。

　　Tegia da vaut建筑一方面能吸引我们的感知，另一方面又通过材料属性和一丝不苟的工艺所形成的某些品质来表达自身。

Tegia da vaut in Domat/Ems

The design was for a space to rise up in the forest of Domat/Ems for the observation of natural and cultural sensory phenomena. The forest had always been a place for unique experiences, both familiar and uncanny. Sheltered in the woods, the "Tegia da vaut" represents a kind of musical instrument that brings our senses into step with nature. The first aspect in developing a concept for "Tegia da vaut" involved gathering materials for the new building from the immediate surroundings. This was informed by a conviction that locally sourced materials should be at the center of a differentiated and site-specific building culture. Diversity can only arise and be preserved by setting boundaries.

In devising "Tegia da vaut", the intention was to use as much of the tree as possible. This vision meant that no components were unusable. The different sorts of timber would buttress, support, interlink and divide. Here, the wood appears layered and heavy; there, it is woven to bring out lightness. The goal was to create a total spatial atmosphere. The acoustics of the insulating sheep wool and the "concentrated" warmth of the soapstone oven particularly influence the lodge's unique overall impression.

"Tegia da vaut" engages our senses on the one hand, and on the other, expresses itself through certain qualities evolving from its material properties and painstaking craft.

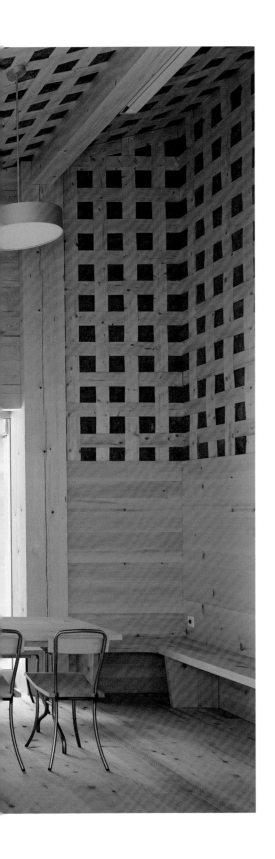
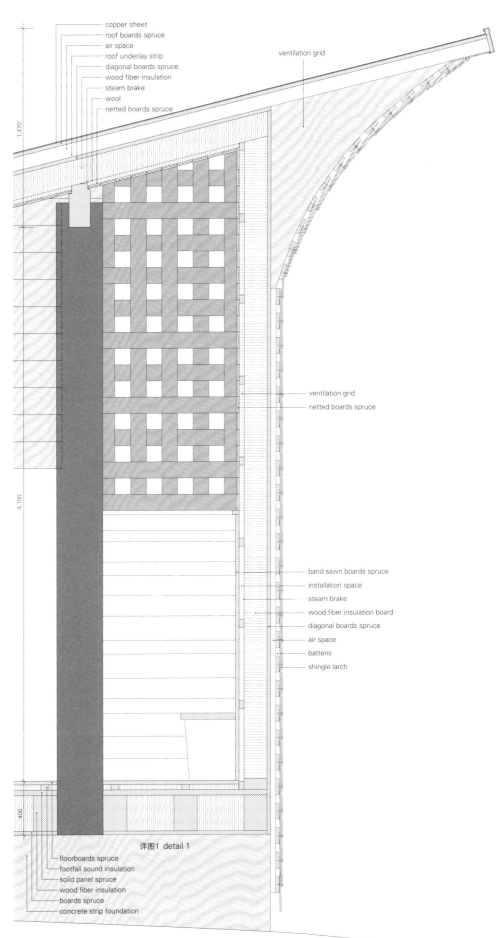

详图1 detail 1

探索瑞士建筑的异曲同工之妙 Exploring the Divergent Forces of Swiss Architecture

巴尔斯塔尔某住宅
Pascal Flammer

该木屋能以不同的方式感知周围的景色。主要分为两个楼层：一层低于地面75cm，二层高出地面150cm。

一层只由一间大型供全家人起居的房间构成，且天花板很低，通过连续的窗户与大自然融为一体，而二层的空间布局则与一层恰恰相反。二层空间被分为面积相同的四个房间，层高达到6m，高度决定了空间。人们透过大型窗户可以将麦田景致尽收眼底。和一层直接与自然亲密接触的特色相比，楼上则是用来观察大自然——一种远距离思考的活动的地方。

项目名称：House in Balsthal
地点：Balsthal in the Protected Area of Jura, Canton of Solothurn
建筑师：Pascal Flammer
合作：Yuu César Barreyre
用地面积：964m²
总建筑面积：186m²
有效楼层面积：337m²
设计时间：2005—2007
施工时间：2007—2014
摄影师：©Ioana Marinescu

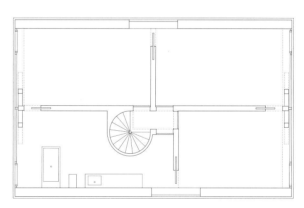

二层 second floor

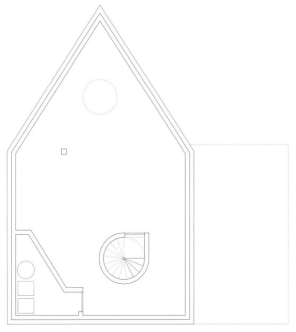

地下一层 first floor below ground

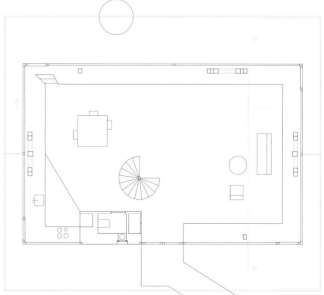

一层 first floor

House in Balsthal

This timber house is about different ways of perceiving the landscape surrounding it. There are two principal floors: one set 75cm below the earth, one 150cm above. The ground floor consists of one single family room with a noticeably low horizontal ceiling. In this space, there is a physical connection with the nature outside the continuous windows. The space above is inverse. This floor is divided into four equal rooms with 6m high ceilings. The height defines the space. Large windows open to composed views of the wheat field. Whereas the ground floor is about connecting with the nature of the context, the floor above is about observing the nature – a more distant and cerebral activity.

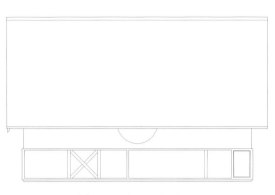

东南立面 south-east elevation

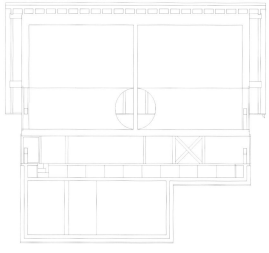

A-A' 剖面图 section A-A'

B-B' 剖面图 section B-B'

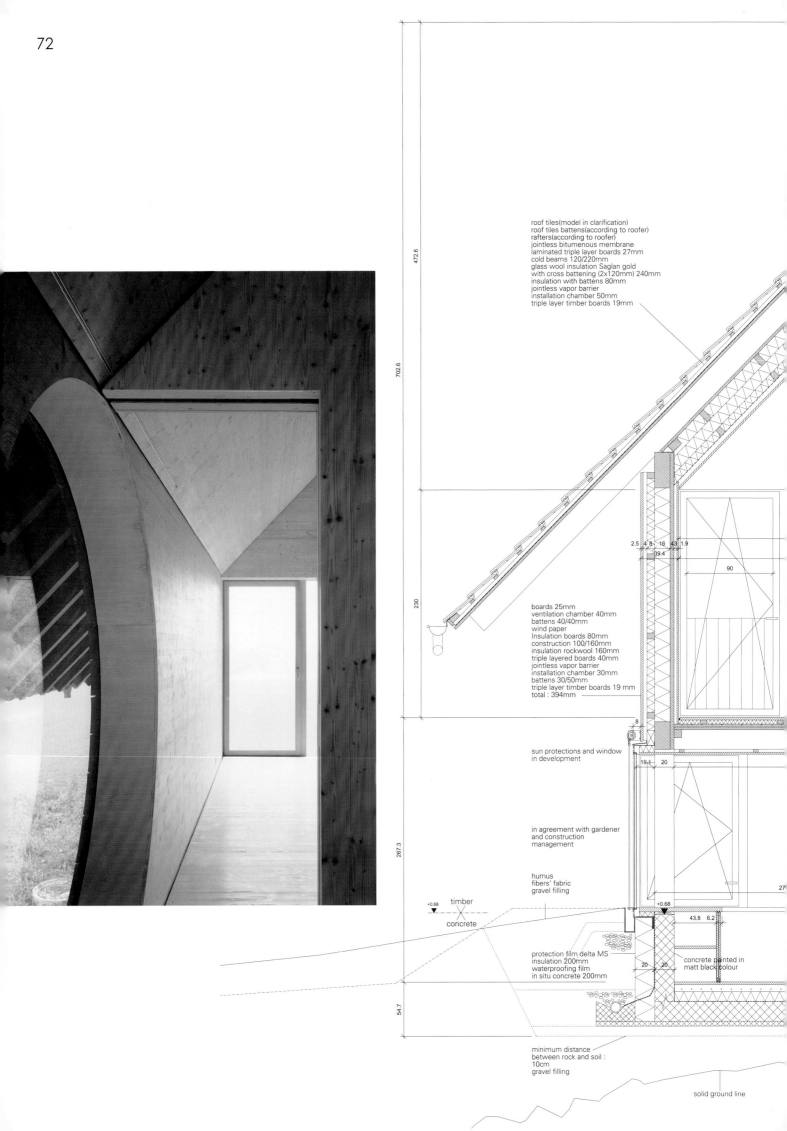

roof tiles(model in clarification)
roof tiles battens(according to roofer)
rafters(according to roofer)
jointless bitumenous membrane
laminated triple layer boards 27mm
cold beams 120/220mm
glass wool insulation Saglan gold
with cross battening (2x120mm) 240mm
insulation with battens 80mm
jointless vapor barrier
installation chamber 50mm
triple layer timber boards 19mm

boards 25mm
ventilation chamber 40mm
battens 40/40mm
wind paper
Insulation boards 80mm
construction 100/160mm
insulation rockwool 160mm
triple layered boards 40mm
jointless vapor barrier
installation chamber 30mm
battens 30/50mm
triple layer timber boards 19 mm
total : 394mm

sun protections and window
in development

in agreement with gardener
and construction
management

humus
fibers' fabric
gravel filling

timber
concrete

protection film delta MS
insulation 200mm
waterproofing film
in situ concrete 200mm

concrete painted in
matt black colour

minimum distance
between rock and soil :
10cm
gravel filling

solid ground line

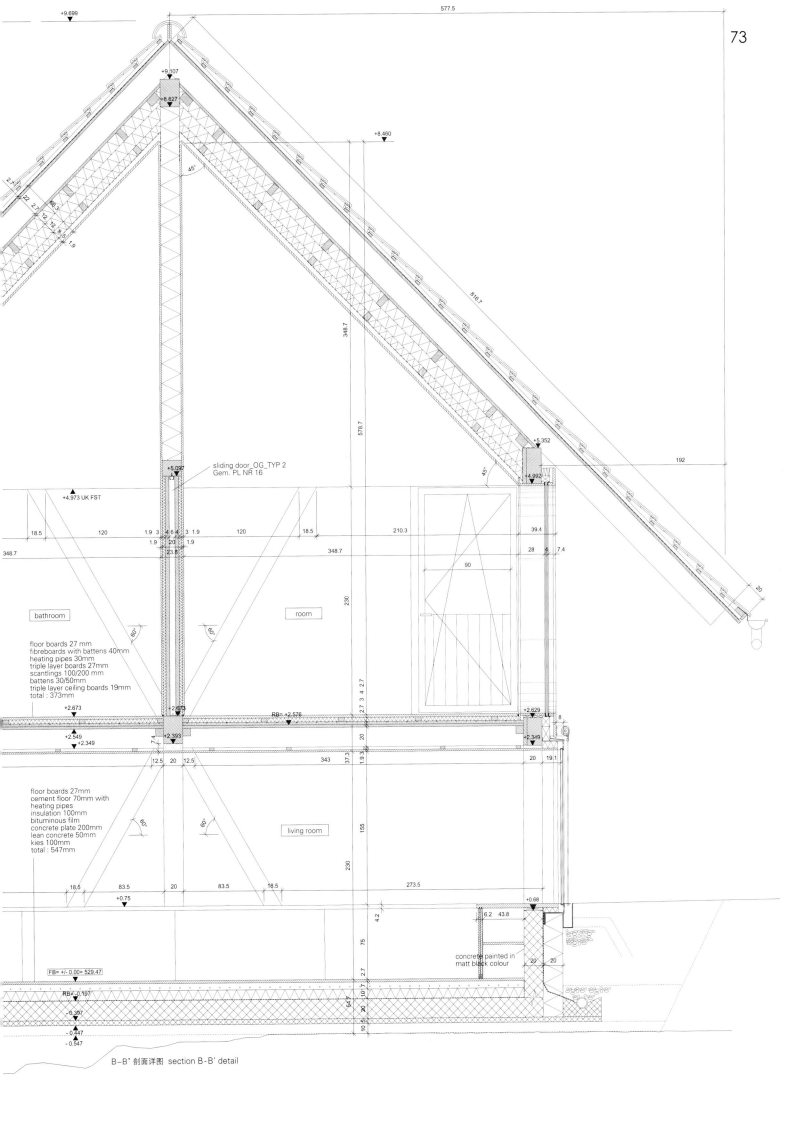

Rotsee三角洲自然景观独特，嵌在两山之间，湖面非常平静。因为这里是一处理想的举办划船比赛的地方，因此在划船手中享有"湖神"的美誉。对新终点站的要求多样并且复杂。基于其功能和周边景观考虑，设计的主要任务就是容易识别。通过把空间单元叠加在一起，一个直立的建筑体量在此形成，并在宽阔的Rotsee湖面上形成一个参考点。尽管这个建筑体量相当大，但是由于三层之间微微错开，让它看起来精巧但不稳固。

终点站项目是Naturarena Rotsee区域发展第一阶段的一部分。赛艇运动中心开幕的时间为2016年7月。终点站和未来的赛艇运动中心通过相同的材质、构造和美学主题而成为一个建筑整体。

三层高的预制木结构位于水面之上的柱状支撑的混凝土平台上。静态的混凝土平台为人们提供了从水上和岸边到达终点站的路径。终点站的背面有楼梯，立面并不突出，因此混凝土结构使建筑贴近湖岸。这展现了终点站建筑的双重特色，一方面具有很强的功能性，另一方面作为湖中的一个雕塑。

该建筑只会在划船比赛期间使用，也就是每年夏季使用三个星期。多数时候都是关闭的，静静地矗立在水面上，所有的百叶窗都关上，成为一处神秘的雕塑般的房子。这种每年都要发生的变化正是设计终点站这个项目所面临的宏伟挑战。

为了在实用功能和雕塑般的美感之间找到平衡，对位于风景如画景观中突出位置的终点站项目而言，其建筑表现是必要的。设计师强调了终点站关闭、滑动的百叶窗缩回后给人的美感。大型滑动百叶窗使建筑外立面具有浮雕感，让终点站看起来可塑性很强，像一座住宅一样。

终点站的外观类似于一件古典雕塑，观察者的位置不同所看到的终点站也不同，完全融于周围的自然景观之中，受日月更替、四季变化的影响。然而，尽管外观看起来很抽象，但其本身非常容易识别，很容易让人知道这栋建筑是划艇比赛时使用的，并且很好地展示了建筑的功能、终点站周围的通道和堆叠的单元。

OK-FISA（国际赛艇联合会）、计时裁判区和解说区三个功能单元与终点线排列在一条轴线上，一个位于另一个之上。较短的建筑立面面向终点线，较长的建筑立面面向终点区域，表明这里是运动赛事的终点。

终点站建筑是木质结构，为了节省时间和成本，由预制构件组成。外立面使用的木头是经过特殊处理的松木，来自再生丛林。木材经过创新方法的处理，使用压力、热和醋酸，使木头减少吸水能力，不变形，更加稳定耐用。

Rotsee Finish Tower

Rotsee-Delta is a unique landscape, embedded in between two hill chains and the lake is very calm. Due to its ideal character for rowing regattas the lake is called the "Lake of Gods" amongst rowers. The requirements for the new finish tower were various and complex. Based on its function and the surrounding landscape the main aim was to create identity. By stacking the spacial units, the vertical volume achieves a point of reference on the wide horizontal plane of the Rotsee. By subtle offsets of the three levels, the volume seems fragile and delicate, despite it's considerable volume.

The finish tower is part of the first phase of the Naturarena Rotsee area development. The opening of the rowing center is July 2016. The finish tower and the future rowing center will form one architectural ensemble, perceivable by the mutual materialization, constructive and aesthetic themes.

The three story high, prefabricated wood construction is carried by a pillared concrete platform above the water level. The statical concrete platform provides access to the tower from the water and the shore. In combination with the stairway on the rear, but no less prominent facade of the building, the concrete structure anchors the building close to the lakeshore. This allegorizes the hybrid character of the building,

Rotsee终点站

Andreas Fuhrimann Gabrielle Hächler Architekten

西北立面
north-west elevation

东南立面
south-east elevation

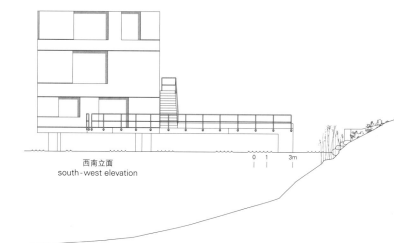

西南立面
south-west elevation

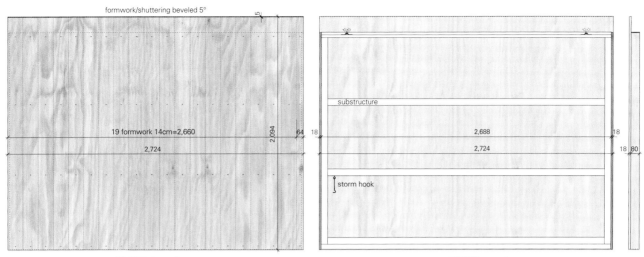

前侧视野 front view　　后侧视野 rear view

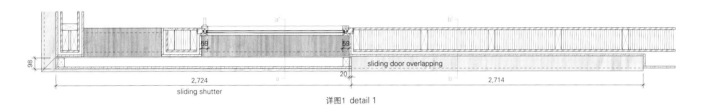

详图1 detail 1

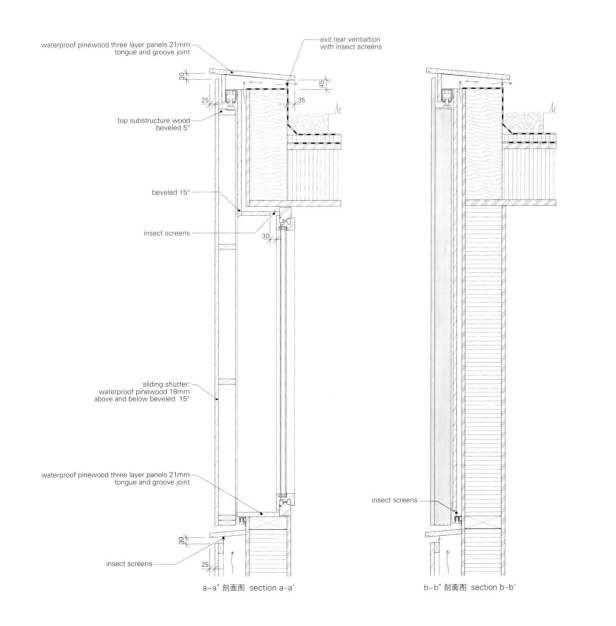

a-a' 剖面图 section a-a'　　b-b' 剖面图 section b-b'

being a functional active building on one side and a sculpture in the lake on the other.

While the building is in use only during the rowing regattas, three weeks every summer, it usually remains closed and stands still on the reflecting water surface, transformed in an enigmatic sculpture-like house, with its shutters closed. This metamorphosis taking place every year was the ambitious challenge in designing the finish tower.

Architectural manifestation for this prominently situated finish tower in the picturesque landscape is necessary in order to find the balance between the practical functional and the sculptural-aesthetic requirements. The aesthetic impression of the tower is emphasized once the building is closed and the sliding shutters are retracted. The large-sized sliding shutters give the facade a relief-like expression and let the tower appear plastic and house related.

Similar to a classical sculpture the tower changes its appearance depending on the position of the observer and blends into the surrounding natural landscape, influenced by the constantly changing days and seasons. The abstract form has a strong recognition value, and therefore conveys identity for the rowing sport, illustrating the function of the building, the context related access of the tower and the stacked units.

The functional units OK-FISA, Jury-Timing and Event-Speaker are axially arranged with the finish line, one above the other. Whilst the shorter facade is pointing towards the finish line, the longer facade is facing towards the finish area indicating the end of the sports ground.

The wooden construction of the finish tower consists of prefabricated elements, in order to build cost- and time efficiency. The wood used for the facade is a specially treated pinewood, from sustainable forests. A innovative method using pressure, heat and acetic acid brings the wood to reaction so that the ability of absorbing water can be reduced essentially, making the wood dimensionally stable and extremely durable.

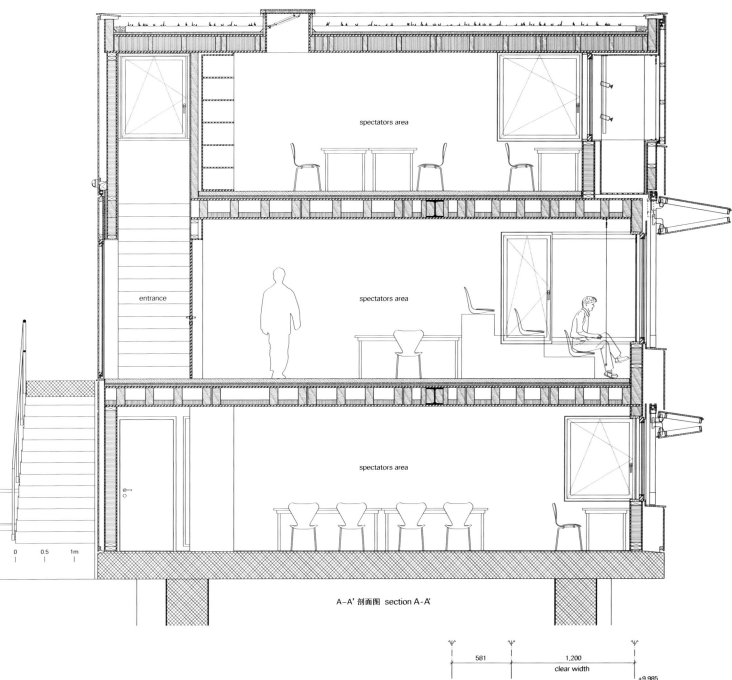

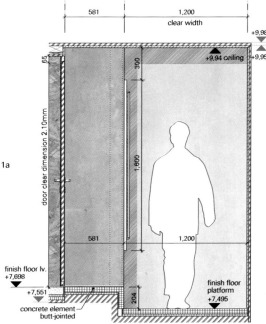

项目名称：Zielturm am Rotsee
地点：Rotsee, Lucerne
建筑师：Andreas Fuhrimann Gabrielle Hächler Architekten
合作：Daniel Stankowski, Carlo Fumarola
规划师：Schärli Architekten AG _ Constr. supervisor, Berchtold + Eicher AG _ Ing., Lauber Ingenieure für Holzbau & Bauwerkserhalt _ Wood structural eng., 1a Holzbau Hunkeler _ wood construction
客户：Naturarena Rotsee
总建筑面积：123m²
设计时间：2012.5—2013.3
施工时间：2012.12—2013.5
竞赛人员：Andreas Fuhrimann, Gabrielle Hächler, Carlo Fumarola, Lukas Schlatter, Andrej Zouev
造价：CHF 1,3 Mio.
摄影师：©Valentin Jeck (courtesy of the architect)

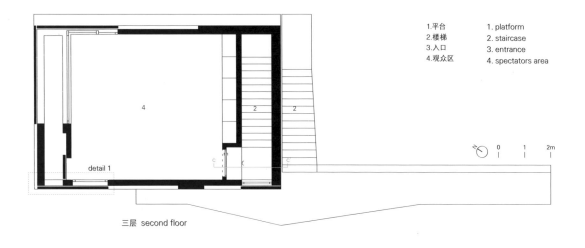

三层 second floor

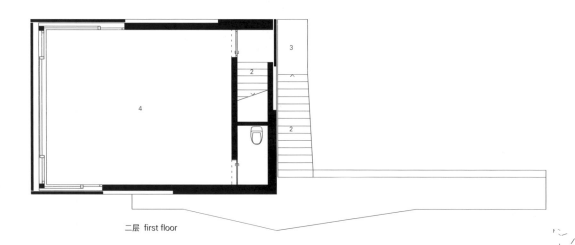

二层 first floor

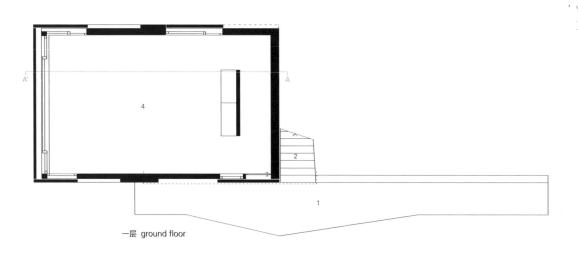

一层 ground floor

1.平台	1. platform
2.楼梯	2. staircase
3.入口	3. entrance
4.观众区	4. spectators area

探索瑞士建筑的异曲同工之妙 Exploring the Divergent Forces of Swiss Architecture

瑞士鸟类研究所游客中心

mlzd

瑞士鸟类研究所新游客中心已在森帕赫湖畔建成，其作用是向公众介绍鸟类研究所的工作成果以及瑞士的鸟类信息，包括一处新展览空间，以及一个治疗生病或受伤的鸟的护理中心。

原来的游客中心位于离湖很近的保护区内，被新设施取代。现在，新设施位于该区域的外面，即湖泊和去往森帕赫村庄的道路之间一个形状不规则的地块之上，因此决定了新游客中心有棱有角的平面布局。

活跃的、四通八达的线路引导游客游览展区和整个湖边景点。紧凑的多边形建筑结构与乡村、湖面相映成趣。其中一个建筑体量是展览中心，展览空间非常灵活，专门用于展示鸟类和研究所的工作成果。

另一栋建筑的一层设有一座礼堂和一个护理中心。其上面的一层有学习空间，顶层有装饰简约的卧室，供在研究所工作的寄宿实习生使用。两栋建筑之间的区域被设计成宽敞的大厅，从这里可以到达展览中心的所有地方。大厅里设有前台以及通往两侧不同设施的入口。在大厅的另一端展示了一个大鸟笼，成为室内和室外的过渡区，逐渐地由室内过渡到室外。大鸟笼旁边是一处休闲座位区，游客可以坐在这里看看鸟儿，也可以通过旁边的窗户向外看湖泊。

该建筑的自支撑的外壳由夯土实心墙构成。这两座建筑共用了512个夯土构件，与混凝土一起，形成建筑的外墙。每个夯土构件平均重达3.4吨。同其他地区的夯土墙设计一样，泥土的暖意营造出平静、柔和的氛围。

另外，简单的木质结构使游客中心的外观富有自己的特色，并合乎可持续发展的理念。夯土有助于降低能耗。另外一些清洁技术，包括通过地热探针和一个热泵来采暖，都有助于该建筑达到所要求的节省能源的标准。同样，利用屋顶光伏发电系统发电和回收利用雨水用于卫生间也是如此。结合低能耗能源的使用，所有的节能措施保证了该建筑获得Minergie-P-ECO证书。

Swiss Ornithological Institute Visitor Center

The new visitor center at the Swiss Ornithological Institute has been built on the banks of Lake Sempach. Its function is to provide the public with more information about the work of the Ornithological Institute and about birdlife in Switzerland accommodating a new exhibition space as well as a care center for treating sick or injured birds.

The new facility replaces an existing center that was situated in a protected area close to the lake. It is now located outside of this area on an irregularly shaped plot between the lake and a road leading to the village of Sempach, which informed its angular plan.

A dynamic network of routes guides visitors through the exhibition and the entire lakeside site. The compact, polygon structures have been positioned so that they interact with the countryside and the lake. One of the building's volumes contains a flexible exhibition space dedicated to birdlife and the work of the institute.

The other houses an auditorium and a care center on the ground floor, with a learning space on the above level and minimally decorated bedrooms for boarding interns who spend time working at the institute on the top floor. The

东立面 east elevation

西立面 west elevation

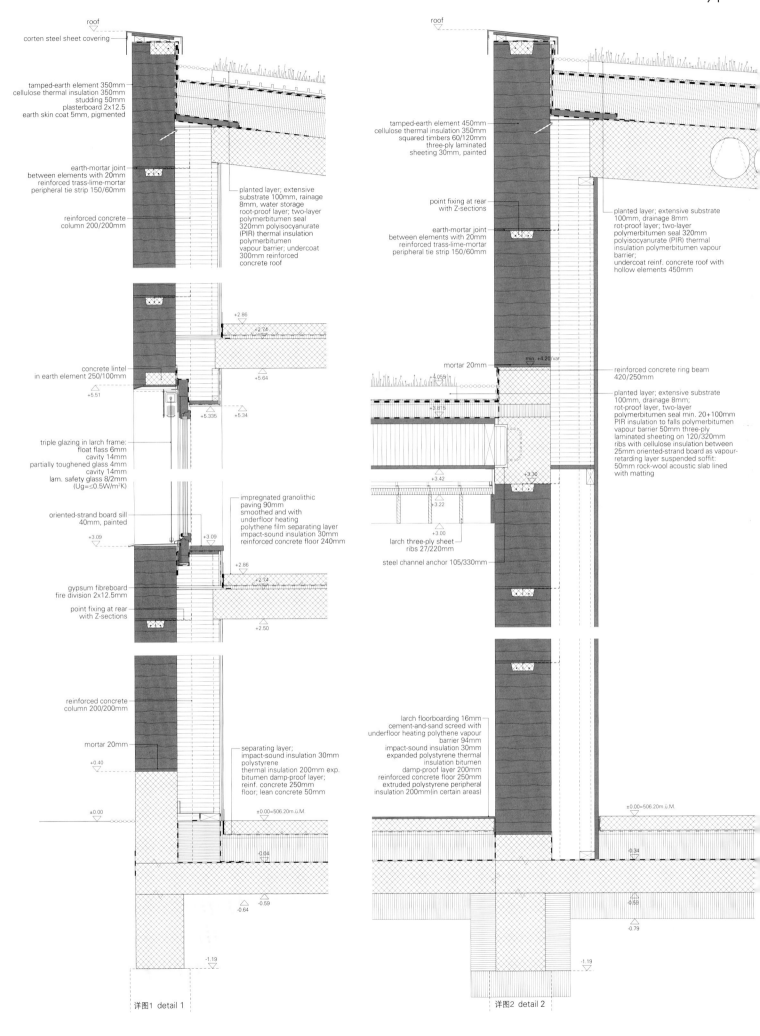

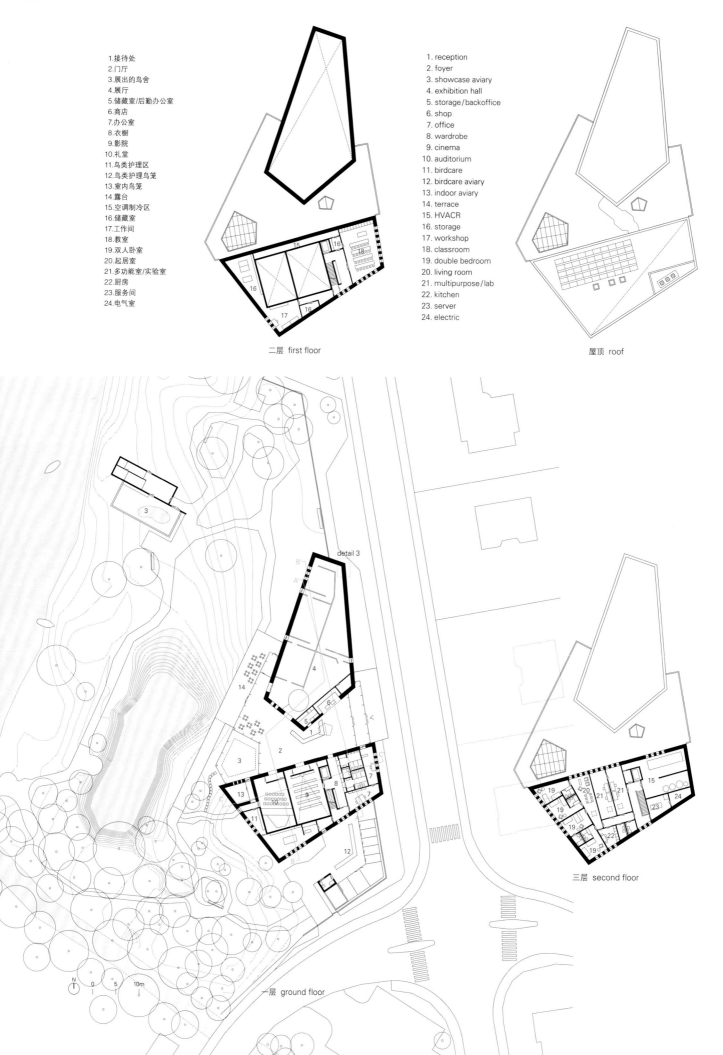

area between the buildings has been made into a spacious foyer, from where there is access to all parts of the exhibition. It features a reception desk and entrances to the various facilities on either side. The showcase aviary at the end of the foyer offers a gradual, fleeting transition between the inside and outside worlds. A casual seating area positioned next to the aviary enables visitors to observe birds in this enclosure, or look out on the lake through the adjacent windows.

The building's self-supporting outer shell is comprised of solid walls made of rammed earth. The two buildings are comprised of 512 earth elements which sit alongside the concrete. The elements have an average weight of 3.4 tonnes. As elsewhere the earth's quiet warmth makes for a peaceful, even subdued, atmosphere.

With the addition of simple timber constructions, they give the visitor center its characteristic appearance and do justice to the sustainable concept. While the rammed earth contributes to the low energy footprint, clean tech features, including heating through geothermal probes and a heat pump helping the building meet the required energy standard, as did a rooftop photovoltaic system producing electricity and rainwater recycling for the toilets. Taken together with the low embodied (grey) energy this suite of kit ensured the building gained the Minergie-P-ECO certificate.

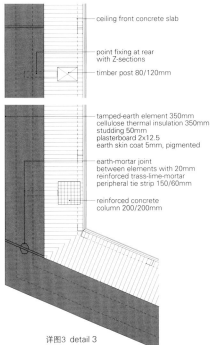

- ceiling front concrete slab
- point fixing at rear with Z-sections
- timber post 80/120mm

- tamped-earth element 350mm
  cellulose thermal insulation 350mm
  studding 50mm
  plasterboard 2x12.5
  earth skin coat 5mm, pigmented
- earth-mortar joint between elements with 20mm reinforced trass-lime-mortar peripheral tie strip 150/60mm
- reinforced concrete column 200/200mm

详图3 detail 3

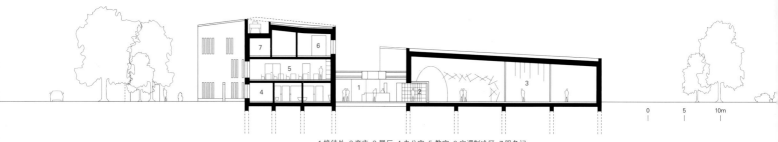

1.接待处 2.商店 3.展厅 4.办公室 5.教室 6.空调制冷区 7.服务间
1. reception 2. shop 3. exhibition hall 4. office 5. classroom 6. HVACR 7. server
A-A' 剖面图 section A-A'

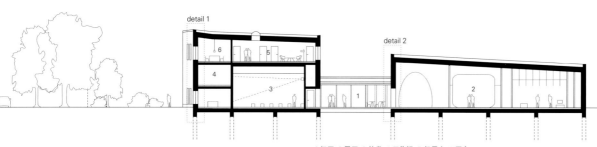

1.门厅 2.展厅 3.礼堂 4.工作间 5.起居室 6.厨房
1. foyer 2. exhibition hall 3. auditorium 4. workshop 5. living room 6. kitchen
B-B' 剖面图 section B-B'

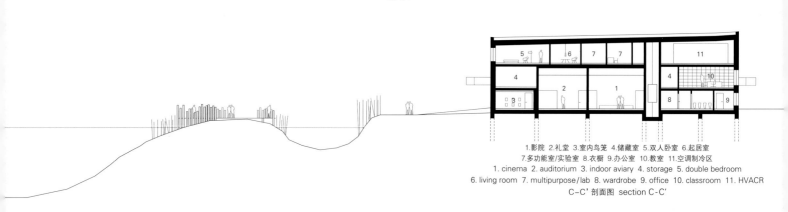

1.影院 2.礼堂 3.室内鸟笼 4.储藏室 5.双人卧室 6.起居室
7.多功能室/实验室 8.衣橱 9.办公室 10.教室 11.空调制冷区
1. cinema 2. auditorium 3. indoor aviary 4. storage 5. double bedroom
6. living room 7. multipurpose/lab 8. wardrobe 9. office 10. classroom 11. HVACR
C-C' 剖面图 section C-C'

项目名称：Visitor Center at the Swiss Ornithological Institute
地点：Luzernerstrasse 6, 6204 Sempach, Switzerland
建筑师：mlzd
项目团队：Claude Marbach, Julia Wurst, Pat Tanner, Daniele Di Giacinto, Roman Lehmann, Amelie Braun, Katharina Kleczka, Marlies Rosenberger, Regina Tadorian, Johannes Weisser, Samuel Wespe, Miriam Zenk
景观建筑师：Fontana Landschaftsarchitektur GmbH, Basel
客户：Swiss Ornithological Institute
功能：visitor center with exhibition, bird care center, classroom, accommodation
用地面积：10,113m² / 总建筑面积：1,400m² / 有效楼层面积：2,010m²
设计时间：2010 / 竣工时间：2015
摄影师：©Alexander Jaquemet(courtesy of the architect)

苏黎世瑞士国家博物馆扩建
Christ & Gantenbein

苏黎世国家博物馆扩建项目的开张庆典于2016年7月举行,是对建筑师古斯塔夫·金1898年设计的原博物馆建筑的补充。新馆位于旧馆的侧面,毗邻Platzspitz公园。新旧建筑物相互直接耦合,形成既有建筑特色又有城市特色的建筑整体。历史和现代建筑元素相视而对,相得益彰。新建筑直接将场地内原有的特色融入其中。

新建筑的布局将这个历史悠久的公园中的树木和道路也纳入其中,并且老建筑富有特色的屋顶景观也成为新建筑的主题。新建筑富有表现力的折叠式屋顶设计可以被理解为是对建筑师古斯塔夫·金清楚明了的历史主义的当代阐释。因此说,如果没有旧建筑,新建筑的风格是标新立异的,但无疑是非常现代的。

从建筑上来说,整座博物馆由两个完全不同的部分组成:一个是优雅而历史悠久的古老建筑,呈开放的U形布局;另一个是富有雕塑感的新翼楼,将原来旧建筑的U形开口封闭,从而新旧建筑之间形成连续的通路。新翼楼包括灵活多样的展览空间、一座图书馆和一座可以举办公共活动的宽敞礼堂。新建筑的中心主题是桥。桥体横跨以水池为主要特色的大型空间(其景观于五月完成),将新的内部庭院和公园连为一体。这座醒目的大桥将一直延续到室内,演变成宽敞大气的楼梯,一直通到最大展览区,成为礼堂的后殿。

尽管新老建筑有所不同,但相似之处和结构共享的特性是显而易见的,这使新旧建筑成为统一的整体。新楼的墙体厚达80cm,与19世纪旧建筑坚实的石墙呼应,同时可以满足Minergie-P环保标准中的高保温要求。新建筑使用的专门开发的凝灰岩混凝土与旧建筑的凝灰岩外立面一致,而新建筑中采用的抛光混凝土地面是对旧建筑中装饰的水磨石地面的现代诠释。

新博物馆的内部主要采用混凝土材料。结合一些暴露在天花板上的技术构件,营造了一种非常工业化的氛围,结实坚固,宽敞明亮,可以用来举行多种多样的展览和展示。这座苏黎世国家博物馆的新空间被视为博物馆工厂大厅,兼具保护性和实验性。

根据国家遗产文物标准,新馆施工阶段还包括历史建筑的改装措施(地震和消防安全)以及针对旧建筑的大面积翻修。博物馆主入口被移到利马特河一侧的新旧馆交界处,这一侧原本坐落着一座美术学校。所有与游客相关的基础设施,包括大厅、衣帽间、商店和餐厅,同新入口一道被改造。在夏季,餐厅和酒吧将在新设计的博物馆广场提供

露天广场餐饮设施，为这个位于火车总站对面的城市中心增添了新的空间和活力。

现在，一个向公众开放的现代研究中心位于利马特河一侧具有悠久历史的翼楼之上，而博物馆的行政管理部门设在旧博物馆建筑的顶层。

然而，苏黎世国家博物馆结构翻新的最后阶段还没有到来。从2017年至2020年，博物馆具有悠久历史的西翼楼和主塔楼将进行翻修。到2020年，博物馆将再次全面向公众开放，从而开启博物馆展示瑞士历史的一个新的篇章。

Swiss National Museum Extension Zürich

The expansion to the National Museum Zürich, of which the opening celebration took place in July 2016, complements the original museum building of 1898 designed by the architect Gustav Gull. The new wing is located on the side adjoining the Platzspitz Park. The old and new buildings are directly coupled to each other so as to form an architectural and urban ensemble. The historical and modern building elements successfully confront each other. The new building directly incorporates some of the context's existing features into its architecture.

The building's layout accommodates the trees and paths of the historical park, and the characteristic roofscape of the old building sets the theme of the new structure. The expressive folds in the rooftops can be understood as a contemporary interpretation of Gull's articulated Historicism. The new is thus inconceivable without the old, but is nonetheless unmistakably modern.

Architecturally the ensemble consists of two very different aspects: the graceful, historical old building designed in an open U shape and the sculptural character of the new wing that closes off the existing building complex thereby enabling continuous movement through both the old and new sections. The new wing includes flexible exhibition spaces, a library and a spacious auditorium for public events. The central motif of the new building is the bridge. It spans across a wide space featuring a water basin (landscaping was completed in May) that connects the new inner courtyard with the park. The prominent bridge carries over into the interior in the form of a monumental set of stairs leading to the largest exhibition area and as a tribune in the auditorium.

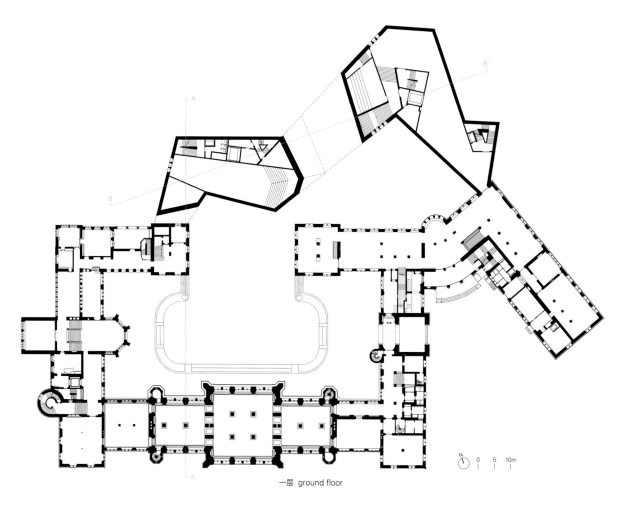

一层 ground floor

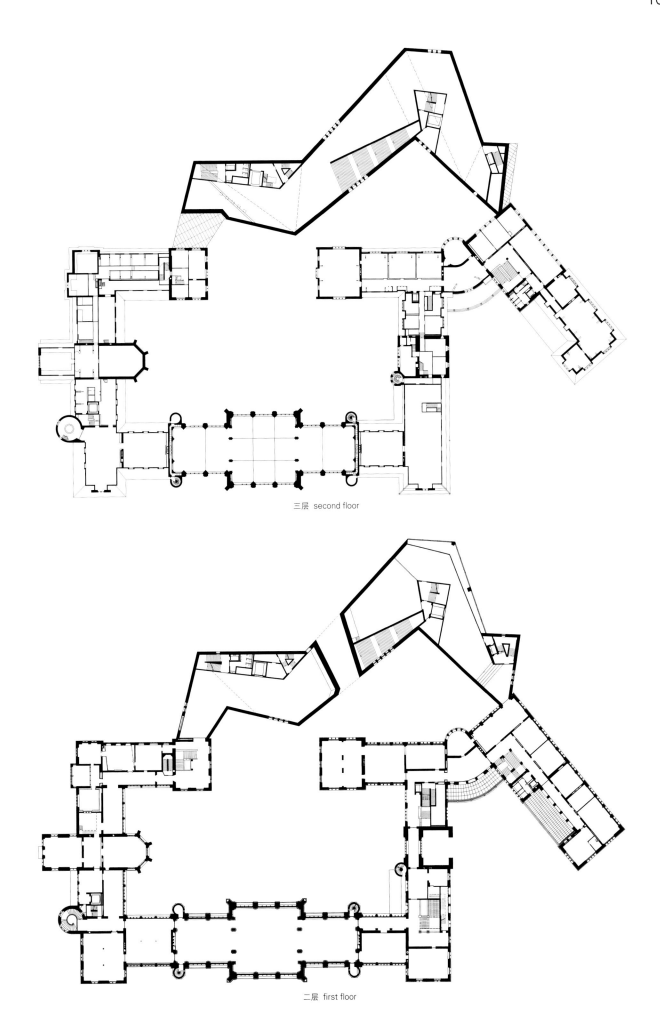

三层 second floor

二层 first floor

As different as the new and old buildings are, their similarities and shared architectural attributes are abundantly apparent and serve to create a unity of old and new. The strong stone walls of the old 19th century building are echoed in the new wing's 80cm thick walls, which fulfil the high thermal insulation requirements of the Minergie-P Eco standard. The tuff concrete developed especially for use in the new wing corresponds to the tuff facade of the old building, and the polished concrete floors in the new wing suggest a modern interpretation of the decorative terrazzo floors in the old building.

Concrete dominates in the interior of the new museum. Combined with the technical elements purposely left exposed on the ceilings, this creates an almost industrial-like atmosphere that is robust, spacious and open to a variety of forms of exhibition and presentation. The new spaces at the National Museum Zürich are conceived as museum factory halls – conservational and at the same time experimental.

The construction phase for the new wing also included conversion measures (earthquake and fire safety) and extensive refurbishing of a large section of the old building according to national heritage criteria. The main entrance to the museum

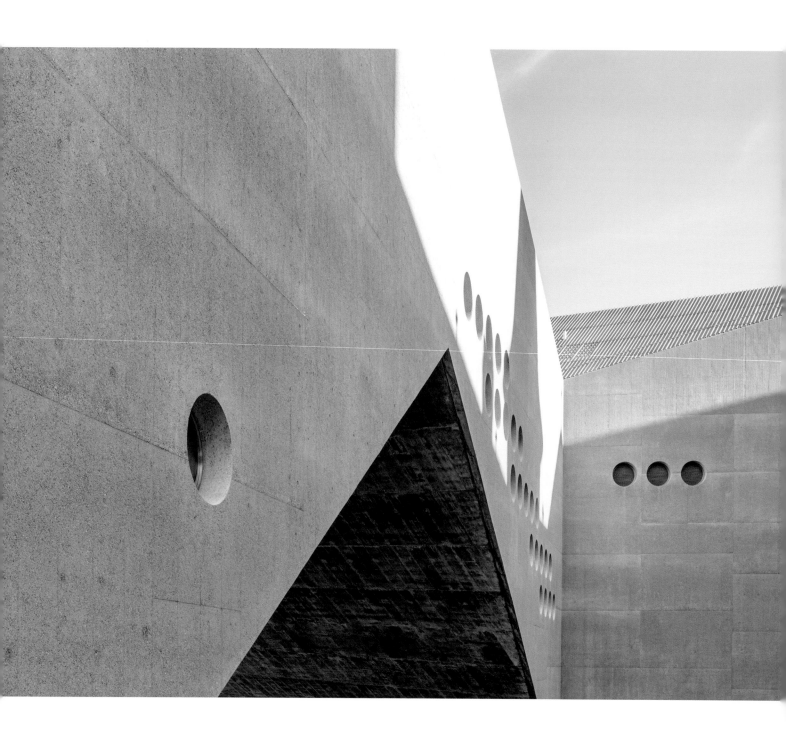

was moved to the spot where the old museum wing meets the wing on the Limmat River side, which originally housed the school of fine arts. Along with the new entrance, the entire visitor's infrastructure, including the foyer, cloak rooms, shop and restaurant, was remodelled. During the summer, the restaurant and bar will offer open-air dining facilities on the newly designed museum plaza, adding a new dimension and vitality to this central urban location, just opposite the main station.

A modern study centre open to the public is now located on the upper levels of the historical Limmat River wing, and the museum's administration is situated on the top floor of the old museum building.

The final phase in the structural refurbishment of the National Museum Zürich is yet to come, however. From 2017 to 2020 the historical west wing and the tower will undergo refurbishing. In 2020 the museum will once again be available to the public in its entirety, thus initiating a new chapter in the museum presentation of Switzerland's history.

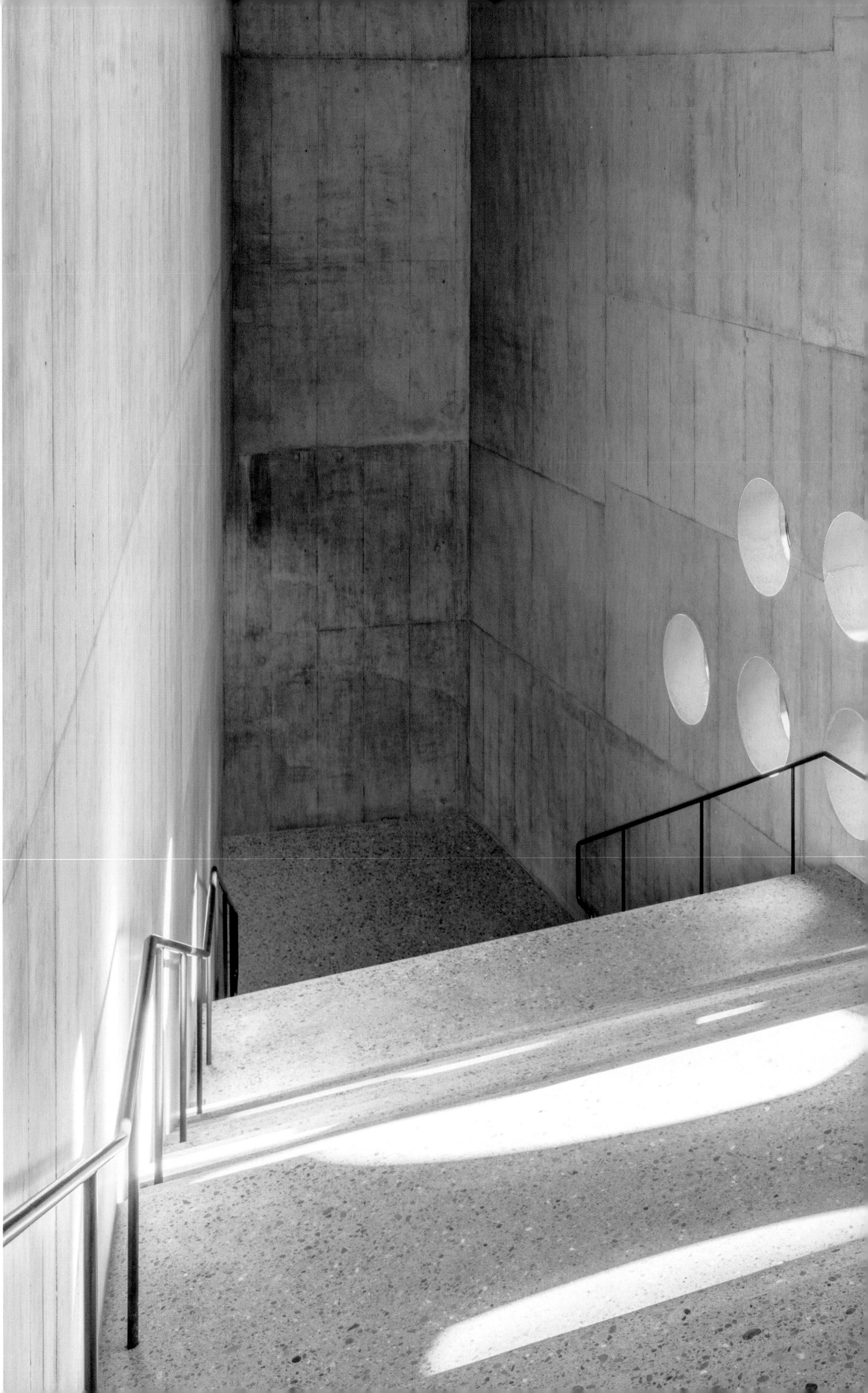

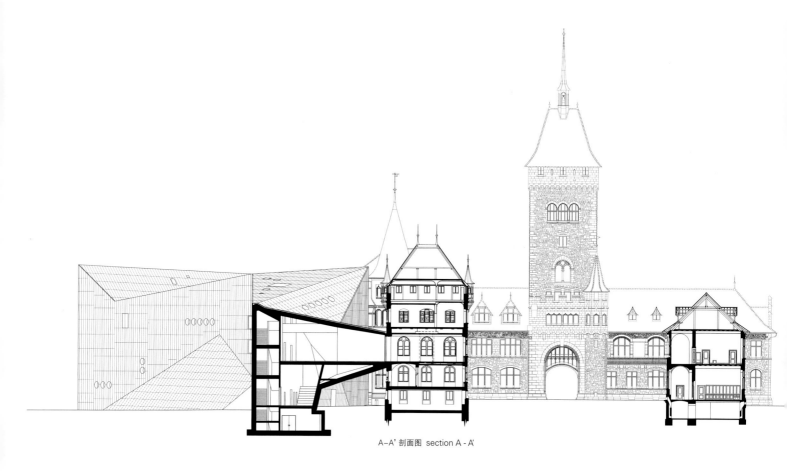

A-A'剖面图 section A - A'

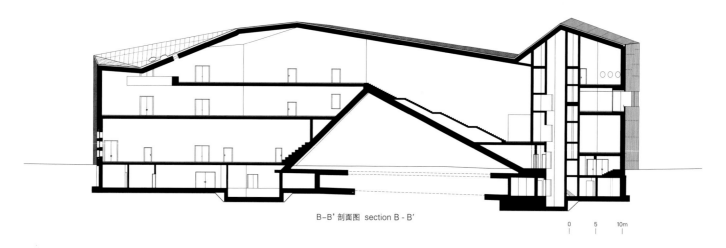

B-B'剖面图 section B - B'

0　5　10m

项目名称：Swiss National Museum Extension Zürich / 地点：Landesmuseum Zürich Museumstr.2 8001 Zürich
建筑师：Christ & Gantenbein / 项目团队：Mona Farag _ overall project manager,
Anna Flückiger _ project manager for school of fine arts wing, Daniel Monheim _ project manager for new wing
业主：Swiss Confederation, represented by the Federal Office for Buildings and Logistics (FOBL)
总承包商：ARGE Generalplaner SLM Proplaning AG / 景观设计：Vogt Landschaftsarchitekten
施工单位：Mona Farag _ overall project manager, Anna Flückiger _ project manager for school of fine arts wing,
Daniel Monheim _ project manager for new wing
场地管理：Peter Guggisberg _ head site manager, Mike Sütterlin (new wing), Matthias Rötzel (old building)
结构工程：APT Ingenieure (old building), WGG Schnetzer Puskas Ingenieure, Proplaning AG (new wing)
用地面积：23,973m² / 总建筑面积：41,800m³ / 有效楼层面积：7,400m²
设计时间：2002 / 施工时间：2016 / 竣工时间：2016 (renovation of west and east wing : 2017—2020)
摄影师：©Roman Keller (courtesy of the architect)

新型社区图书馆

New Commu

通过本章中所展示的项目，我们有机会近距离观察图书馆这一建筑类型的过往、当下和未来，从不同时期探索其存在意义的变化更迭。长久以来，社区图书馆的形象在人们心中一直是墨守成规的形象——老旧的建筑、昏暗的灯光和一排排的厚重书架。其实随着社会和技术环境的发展，如今图书馆的作用已经有了很大的改变，还能不断满足社区要求、风俗和服务方面的需求。新式图书馆逐渐成为地标型建筑，被市民所认同，使文化重新恢复活力。新式图书馆还宣扬一种新的建筑理念，即在维护自然稳定，保护图书馆阅读者和工作人员身体健康的同时，还要满足未来用户的兴趣和需求。值得关注的是，伴随着全球化的进程，各地对自身的身份认同感和存在意义有了更高的需求，由此，诞生了文中所展示的这四个项目，而这些项目也交出了对成功社区图书馆规划的最全面的答卷。

The projects presented within this chapter provide an opportunity to have a closer look at the ever-changing meaning of the library, yesterday, today and tomorrow. Community libraries have come a long way from their former cliché image of an old building with dimmed light and endless rows of heavy wooden bookshelves. Their role is extending rapidly alongside steady major shifts in societal and technological environments, as well as the increasing array of community needs, custom and services. New libraries are progressively seen as pivotal to place making civic identity and cultural revitalisation. They also announce a new mind-set in taking responsibility for the stability of nature, health of library users and staffs, and catering for the needs and interest of future generations of users. And remarkably, the four presented projects all profited from the fact that globalisation has been accompanied by a growing quest for local identity and meaning, allowing these projects to show a comprehensive range of answers for the planning and delivering a successful community library.

Minna no Mori 岐阜媒体中心_Minna no Mori Gifu Media Cosmos/Toyo Ito & Associates, Architects
Saint-Just Saint-Rambert中央图书馆_Central Library in Saint-Just Saint-Rambert/Gautier + Conquet
Constitución公共图书馆_Constitución Public Library/Sebastián Irarrázaval Arquitectos
海岸图书馆_Seashore Library/Vector Architects

规划新式图书馆_Planning the New Library/Tom Van Malderen

规划新式图书馆

一直以来，图书馆都固守着自身的作用和功能，即传承知识和传播文化；然而，有时图书馆也会成为社会的变化的产物。早期的希腊图书馆为图书馆的发展之路奠定了基础，最初，希腊人试图利用图书馆将可以获得的个人记录保留下来，结果却是将人类的全部记忆保留了下来。到了17世纪，图书馆又有了一个新的功能，即学术研究中心。一个世纪后，因为权贵一族才能享受图书馆的进出权，导致了成本高昂和私人定制的图书馆的出现。直到19世纪中叶，才出现目前人们所熟悉的公共图书馆，这也是第一次大众可以自由进入图书馆，获取所需的信息和文献。

在某种意义上，图书馆的变化总是与社会的变革保持同步，不断演化，回顾历史，并展望未来。现代社会随着科学技术的迅速发展而发生着日新月异的变化，这种变化还进而挑战着当今社会的图书馆的变革。现在的图书馆必须面对现实，即当前的环境不再只简单地局限于一个特定的阶段或转折期，而是进行着不可预测的时时刻刻的变化，从而改造成为新型图书馆。因此，图书馆的角色也发生着重大的改变，从之前被动的、书籍储藏为主的形象，转换为一种充满活力、功能设施齐全的面貌。集图书收藏、聚会场所、社区中心、学术中心、人文活动、创新活动、电子信息服务等于一身的综合机构。老旧的图书馆已然将自己从一个储藏室重新打造成为一个多功能平台。通过对本章介绍的建筑的深入了解，我们可以见识到各式变革式的图书馆。Vector建筑师事务所设计的海岸图书馆是度假村中的一部分，设有大型开放阅读区、静思厅、多功能活动室、酒吧和休息区。伊东丰雄设计的Minna no

Planning the New Library

The library has had a sort of constancy in its role and function throughout time: a devotion to sustaining knowledge and culture, despite, and sometimes as a consequence of, the many changes in society. The early Greek libraries paved the way: they were set out to keep a copy of all individual records available at the time, turning themselves into an instrument to preserve the entire memory of humankind. The seventeenth century library added a new role to that space by rematerializing as a centre of scholarship. A century later, costly and private subscription of libraries emerged with limited access to the privileged. The public library as we know it only came to the foreground in the middle of the nineteenth century, providing for the first time widespread and free access to information and literature.

In a sense, libraries always have resonated with changes in society. They always had to evolve and look into the future, not only the past. Today's libraries have yet again been challenged by the dynamics of perpetual change: a change that came along with technology and science developing at an increasing speed. Now, libraries have to live up to the fact that the present environment of unpredictable and perpetual change is probably no longer a phase or a period of transition, but the new reality to be. Their role has changed significantly from passive, collection-based institutions to active, vibrant facilities offering easy access to collections, meeting places, community hubs, learning centres, civic and cultural programs, contents creating activity and services of digital era. These age-old institutions are increasingly recasting themselves as platforms rather than storehouses. Having a closer look at this chapter's projects, we understand that the Seashore Library

Mori岐阜媒体中心则是一座混合式市政大楼，一楼大堂设有种类多样的公共设施：画廊、演讲厅、公共活动中心，二楼则是社区图书馆。

可以说，如今的图书馆正处于剧烈变化时期。为了与不断变化的大环境保持同步，现在的图书馆必须迎合现代信息和社会的需求。现代图书馆不仅要保留传统图书馆的功能，满足人们在信息和交流方面的需求，还要迎合人们的文化、休闲、环境等其他社会需要，不但提供传统的图书馆服务，还要组织非传统的文化社会活动。值得庆幸的是，现代图书馆已经适应了这个不断变化的世界，这个结果比人们十年前预想的要好得多。在这个数字信息时代，图书馆要生存下去，其关键就是要紧紧贴近社区，将图书馆的功能拓展到社区中去，借此产生社会资本，创造融入社会的机会。作为一种建筑类型，图书馆是城市公共空间的重要一环，拥有丰富的历史。这样的地位似乎已经逐渐融入整个社会的规模中，并得到响应。然而奇怪的是，科技其实挽救了很多图书馆——只要图书馆仍然是电脑服务和社区成员继续教育的主要机构。没有这些资源，信息鸿沟将不断扩大，很多社区将急剧退步。现代图书馆是社区的建筑中心，以人为本，赋予居民权利和民主性。图书馆的大门向所有人敞开，使他们进行社会交往和个人冥思，无论他们的背景和社会地位如何。现代图书馆提供给人们的是一种独特的空间组合，既可以是现实社会和现实人们的空间，也可以是连接全球人类和万维网的虚拟空间。

人们已经成为社区图书馆的规划焦点。这一转变体现在，尽管对储存纸质版书籍的空间需求下降，但对能够供人们活动用的公共空间的需求却在不断增加。这样的空间鼓励人们使用图书馆的资源、服务、功能、技术和其他服务。Sebastian Irarrázaval设计的图书馆内沿立面设

by Vector Architects forms a part of vacation compound and houses a large open reading area, a meditation space, a multi-functional activity room, a bar, and a resting area. Also Toyo Ito's Minna no Mori Gifu Media Cosmos is a hybrid municipal building, containing a variety of public amenities on its ground floor, including galleries, a lecture hall, and a public activity centre, as well as the community library upstairs. We can state that today's library is experiencing a period of intense changes. In order to remain relevant in a constantly changing environment, today's library has to respond to modern information and society's needs. It has to correspond not only to informational and communicational needs, but also to cultural, leisure-seeking, environmental and other community needs, to provide traditional library services not only, to but also organise non-traditional, cultural and social activity. Luckily, libraries seem to adapt to this new reality much better than anyone might have anticipated a decade ago. The key to win in this digital era seems to lie in engaging communities, in extending the library's role towards that of the community, and in banking on its ability to generate social capital and to build opportunities for inclusion. As a typology, the library enjoys a rich history as an important component of public space within the city. This position seems to be increasingly established on the scale of the community and community responsiveness. The strange reality is that technology might actually be the saving grace for many libraries, given they remain to date the primary source of computer access and continuing education for members of the community. Without these resources, the digital divide would only increase further, and numerous communities would be set back. Today's libraries are at the heart of the community, putting people first, enabling empowerment and democracy. They are welcoming all and everyone, irrespective of background or status, for social connection or individual contemplation. The contemporary library offers a unique combination of spaces for real time and people, and virtual space for interacting across the globe and

北戴河新区海岸图书馆，中国
Seashore Library in Beidaihe New District, China

置了一连串长椅，这样简单的设计让人们有机会聚在一起互相交流，分享经验。Sebastian Irarrázaval还将室内地面水平高度提高了一米半，这种做法使在这里阅读的人们可以一边读书，一边欣赏广场上的古树，增强了图书馆的公共开放属性。在《C3》第58期中，Douglas Murphy谈到了一种有趣的建筑特色，即将公共空间带入建筑中。实现这种设计的方式有两种，一是创造大面积房间来模拟公共空间，二是将室外空间（类似于建筑的额外房间）定义为建筑的公共空间，即将某些室内特性引入大众领域。

Gautier+Conquet设计的Saint-Just Saint-Rambert中央图书馆很好地诠释了这一特点，中央图书馆坐落于城镇入口，位于一座广场旁边。而在伊东丰雄的设计中，也有一个四周摆放着木质旋转书架的大规模阅览室，书架围合出流线，且阅览室的屋顶呈起伏状，设有漏斗形状的构件，定义了多样化的阅读、休息和学习空间。

这一章中介绍的建筑项目展示了人们对图书馆的主要期望，即给人们提供安全的空间，以及市民可以参与的空间和民主空间。图书馆似乎已经欣然悦纳了Ray Oldenburg在《绝好的地方》中提出的"第三空间"这一概念，将自身融入人们的生活空间内。Ray Oldenburg提出，所谓的"第三空间"是社区生活的"精神支柱"，形成更多的创新互动。图书馆是高度开放的，令人感到温馨和舒适，传统的图书馆则转变为社区的会客厅。随着社会的不断发展，大部分公共领域都逐渐被市场经济和公司所掌控，承受着各方面的压力，图书馆可能是遗留下来的为数不多的独立公共空间了。值得庆幸的是，即便是现在，图书馆仍然处在商业管理以外，是当今鲜有的未被广告入侵的公共场所，它们的存在让人们能够好好地做自己，而不是扮演顾客的角色。

the World Wide Web.
People have become the focus for planning the community library. Whilst space for hardcopy collections is generally decreasing, the required physical space is still on the rise with the need to deliver more space for people. Spaces that welcome people and encourage people to engage with library resources, services, programs, technology and others. In Sebastian Irarrázaval's building, simple benches stretching along the facade give people an opportunity to connect and share experiences. He also lifted the interior a meter and a half above ground level to accompany the reader-user with the view of the old trees in the Plaza and reinforce its public character. With reference to Douglas Murphy's text in C3 no 58, he talks about the interesting architectural quality in bringing the public space into the building, through creating large rooms mimicking public space, or the opposite, the definition of outdoor spaces that become something akin to an extra room of the building, an introduction of a certain domesticity to the public realm.
We see this quality at work in the Central Library in Saint-Just Saint-Rambert from Gautier + Conquet with the adjacent square at the entrance of the town. Also in Toyo Ito's project where a large continuous reading room is organized by a series of wooden spiral bookcases that shape the circulation and roofed over with an undulating roof of light funnels defining multiple reading, resting and studying zones.
The projects in this chapter illustrate a significant desire and intention for libraries to offer safe spaces for people, and spaces that are important for democracy and civic engagement. Libraries seem to have embraced the notion of the "third place", as coined by Ray Oldenburg in his book *The Great Good Place*, and integrated it into their physical space. He argued that "third places" act as "anchors" of community life and foster more creative interaction: they are highly accessible, as well as welcoming and comfortable. The shift happens from libraries of the past to libraries as community

作为以社区为中心的图书馆，其在更丰富的社区对话中起到了重要而独特的作用，包括关于新式建筑的适应能力、气候变化及可持续发展的前景。过去几年里，图书馆成为采用可持续发展设计的新式建筑中最常见的类型之一。环保图书馆强调一种新型责任的思维模式：稳定自然平衡，保护用户和工作人员身体健康，同时满足未来几代用户的需求和趣味。现代图书馆承担着道德责任，事实上，我们建立了全球意识，认识到现在的消费速度、不健康的产品的消耗、不合理的制作过程以及体系，会给整个世界的经济、社会乃至个人带来严重后果，造成不可持续发展的未来。本章中的设计都体现了图书馆的道德观念，并包含了设计中的有效利用能源、水源和其他资源，同时减少浪费、污染和环境破坏的有效途径。同时，除了保护建筑特色和建筑体系，这些建筑的设计还认识到保护参与者健康的重要性（通过对空气和光线等因素加以控制）。

现代图书馆必须要与其所处的自然环境和城市环境相互融合。据《C3》第34期的文章介绍，Aldo Vanini发现，近期的图书馆设计越来越多地利用透明性和通往室外的洞口，从而增强建成环境与自然环境的融合程度。Vector的海岸图书馆项目整体设计把外围环境带入内部空间，逐步赋予大海、阳光和海风三种元素独特的关系，通过它们之间的各种变换，实现从多空间和多功能方面感受生活气息。伊东的团队没有通过明确地分割空间或隐藏空间来实现减少能源消耗的目的，而是通过与自然环境相联通的方式，将岐阜媒体中心打造成舒适的、能源节约型的环境。

另外值得一提的是，本章所展示的所有设计都将木材作为建筑的主要构件及饰面材料。在《C3》第42期，我阐述木建筑的再生，以及

living rooms. With the public domain under growing pressure as a result of society ongoing towards being controlled by the economy and the grip of corporations, libraries might be one of the few truly public places left. Thankfully, even today, they remain largely outside the scrutiny of commercialization and as an increasingly rare public space without the presence of advertisements, they allow people to be people, and not consumers.
Putting its emphasis on the community, the library also plays an important and unique role in wider community conversations about resilience, climate change, and a sustainable future. Over the past several years, libraries have become one of the most common categories of new construction to embrace sustainable design. Green library management emphasizes a new mind-set of taking responsibility for the stability of nature, the health of its users and staffs, and catering for the needs and interest of users of future generations. Today's libraries take on a moral duty. They acknowledge the fact that we have reached a tipping point in global awareness and that our current rate of consumption and use of unhealthy products, processes, and systems are producing a serious, unsustainable impact on the economy, on communities, and on individuals. The projects within this chapter reflect this ethical stance and their approaches extend to include efficient use of energy, water, and other resources, as well as the reduction of waste, pollution and environmental degradation. Along with the physical characteristics and building systems, these buildings recognize the critical importance of protecting the occupant's health by addressing factors such as air and light quality.
The contemporary library has to exist in harmony with its natural and urban environment. In C3 no 34, Aldo Vanini observes that recent examples of libraries have tried to rise to the growing demand for more integration between the built and natural environments, often through the use of transparencies and openings to the outside. Vector's Seashore Library project is designed around pulling the outside in and instilling a unique relationship with the ocean, natural light, and wind; and in doing so, using each of these elements with varying

Constitución公共图书馆，智利
Constitución Public Library, Chile

这一主题是如何利用技术和理念再一次引发人们的兴趣。当然，大部分人感性地认为木材给人以"温暖"的感觉。其实，已有科学依据表明，木材的使用对人的身心健康都是有益的。具有如此生态友好和技术潜质的木质建材，必将在环保建筑和建筑工业中发挥重要的作用。Sebastian Irarrázaval说他们的图书馆建筑最初的设计是使用钢筋混凝土建成一座混合式建筑，但是由于资金问题，最后几乎全部使用木材。他们就地取材，请当地的木匠，不仅丰富了建筑的历史，还将其与自然更紧密地结合在一起。

这些特色设计项目对大量问题进行了回应，并且强调了社区图书馆设计要与更多的目标保持同步。社区获取信息的方式、信息的全球化以及瞬息万变的科技环境都在实体空间的发展中有所体现，满足了各式功能性空间的灵活性需求，来进行学术活动、文化活动和娱乐活动。这样的多样性空间设计实现了在不同时间内举办同一活动，或者是同一时间举办不同活动的功能。图书馆实现了巨大转变，从过去重视图书收藏、设备和相关基础设施的空间设计，转变为如今以人为本，重视社区对外的包容性、用户体验和创新的设计。

degrees to breathe life into the many spaces and functions. Instead of clearly dividing or concealing spaces to reduce the consumption of energy, Toyo Ito's team created with the Gifu Media Cosmos a comfortable and energy-saving environment with a connection to nature.

It is also remarkable that all projects presented in this chapter make use of wood as a major component for the building and its finishes. In C3 no 42, a piece I previously wrote on the recovery of wood and how this renewed interest was stimulated both by technology and ideology. Many people emotionally respond to wood and refer to it as something "warm". The beneficial link between wood and human health, both psychologically and physiologically, has even been scientifically established. With ecological friendliness or technological potential, wood is bound to play a prominent role in green building and construction industry. Sebastian Irarrázaval reports that their library design was initially conceived as a mixed structure with lots of concrete, but due to financial constraints, it ended up being constructed almost entirely of wood. They took advantage of the local supply of wood and carpenters, enriching the project's story and tying it closer to its natural environment.

The featured projects show us a wide range of responses and emphasize that community library planning is undertaken in sync with wider objectives. The way a community accesses information, the globalisation of information, and the rapidly changing technological environment become manifest in the development of physical space, instilling the need for flexibility in spaces to be used for a variety of functions whether it be learning, cultural or recreational activities. This hybridisation permits spaces being used for the same thing at different times or different things at the same time. It documents a massive shift from space planning for collections, equipment and associated infrastructure towards a strong focus on designing for people, for community outcomes, and for experience and innovation. Tom Van Malderen

Minna no Mori岐阜媒体中心

Toyo Ito & Associates, Architects

发现新式能源节约型空间范式,寻求建筑与自然的连接

Minna no Mori岐阜媒体中心是一座综合图书馆,于2015年7月开馆。在这里,人们可以欣赏到岐阜市的标志性风景:金华山和山顶的岐阜城堡。该建筑在日本岐阜市车站北面两公里,且长良川河为图书馆提供了丰富的地下水源。

为了打造一个文化中心,激活城市氛围,岐阜市就图书馆综合体和周围广场的设计方案举行了一次设计大赛。于是,在2011年2月,我们被选定为设计师。建筑分上下两层,主体面积为80m×90m。一层的中心区是玻璃图书档案馆,藏书60万册,周围连续分布着展览厅、多功能大厅以及市民活动和交流中心。

二层是一处无隔断的阅览区,设计了可存书30万册的书架,以及多样式浏览座椅(共910个座位)。

圆形阅览区域和接待处的上方悬挂着名为"环球"的大型半透明伞状吊顶。该构件由聚酯纤维制成,形状像一个倒置的漏斗。11个直径在8m~14m之间的"环球"将产生自上而下的自然风和柔和的漫射光,营造出舒适的室内环境。

微微起伏的木屋顶由符合室内尺度的木材制成,规格为120mm×20mm。轻薄的木板条彼此堆叠,在三个方向延伸。"环球"吊顶位于起伏屋顶的顶点,并与之相映成辉。屋顶被周围连绵的山脉环抱,为城市增添了一派庄严且和谐的景象。

场地种植了很多树木,尤其是建筑的西面,有长达240m的树林走廊,南面则有一座45m宽的广场。二层的三个露台都有自己的特色,营造出一种舒适放松的环境。随着"环球"的设计将自然能源引入建筑,并且利用地下水来为地面供热和制冷,同时利用屋顶的太阳能光板,所有因素结合在一起,形成一个平衡机制。这些建筑设计使其与20世纪90年代同样规模的建筑相比,预期会减少50%的能源消耗。

不同于利用明确的空间布局或者隐藏空间的做法来减少能源消耗,我们创造了一处舒适的能源节约型环境,同时追求建筑与自然的

结合，让人们在空间内感受空间的活力，在空间内享受，这是日本建筑长期追求的主题。

用户可以享受空间，如同身处城市景观内，并且在这样的空间中寻求人与人之间新的联系。

Minna no Mori Gifu Media Cosmos
The discovery of a new energy-saving spatial model in pursuit of connecting with nature

Minna no Mori Gifu Media Cosmos is a library complex, opened in July 2015. The site has a symbolic view of the city: Mount Kinka with Gifu Castle on its top. It is located 2km north from JR Gifu Station and has abundant subsoil water from Nagara River.

In order to create a new cultural centre and to activate the city, there was a design competition for a complex and plaza around it. We were selected as the designer in February 2011. The building is made up of two stories, about 80m x 90m in dimension. The first floor has a glass book archive that accommodates 600,000 books, located in the middle of the large plan. An exhibition gallery, a multi-purpose hall and a civic activity and exchange centre are placed consecutively around the archive.

The second floor is one large reading area with no walls, and has shelves for 300,000 books as well as various kinds of browsing seats (910 seats in total).

Large translucent umbrellas named "Globe", hang above the

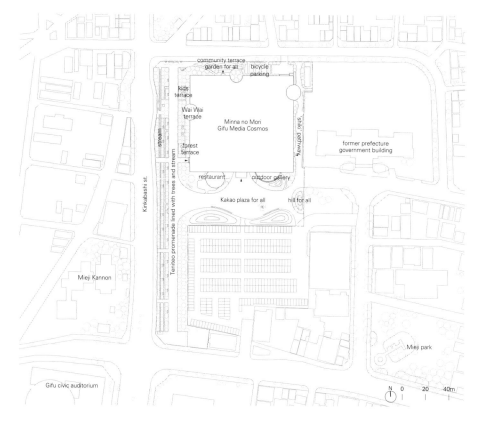

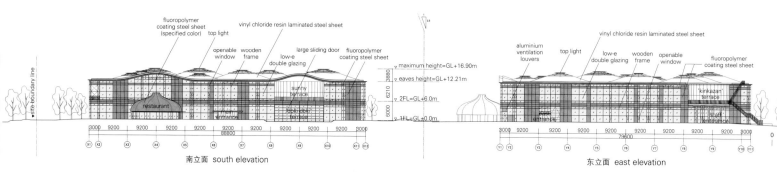

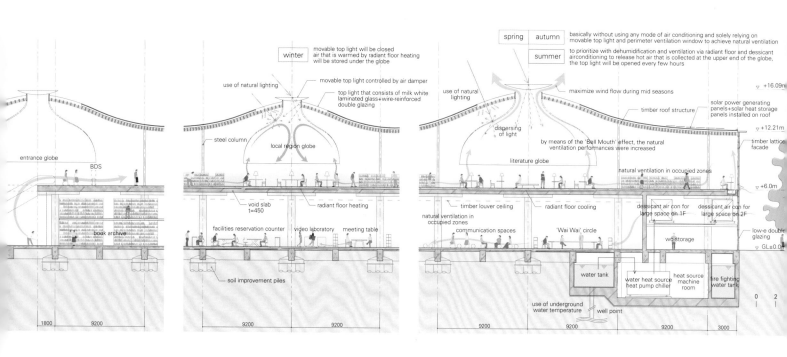

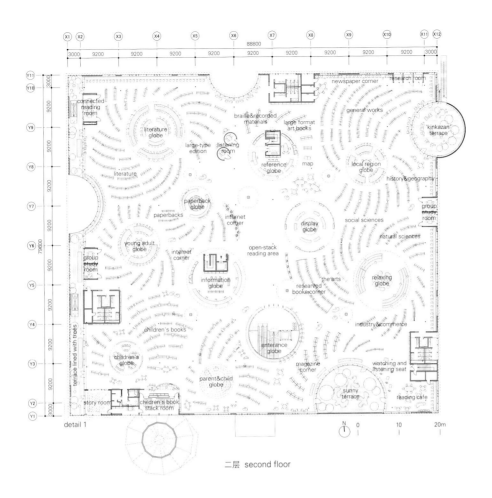

二层 second floor

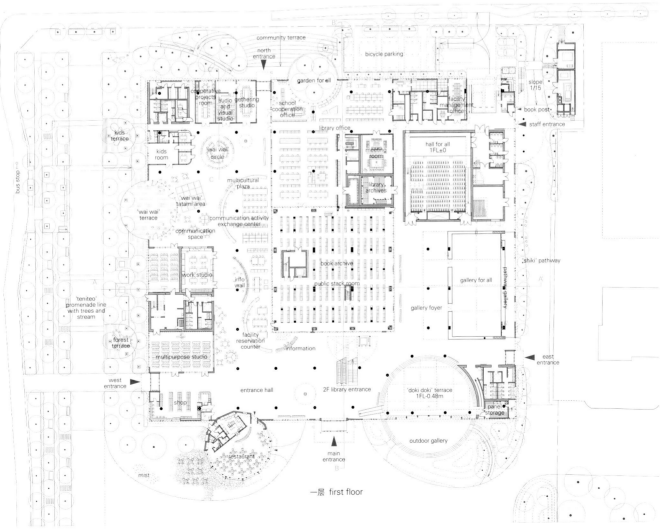

一层 first floor

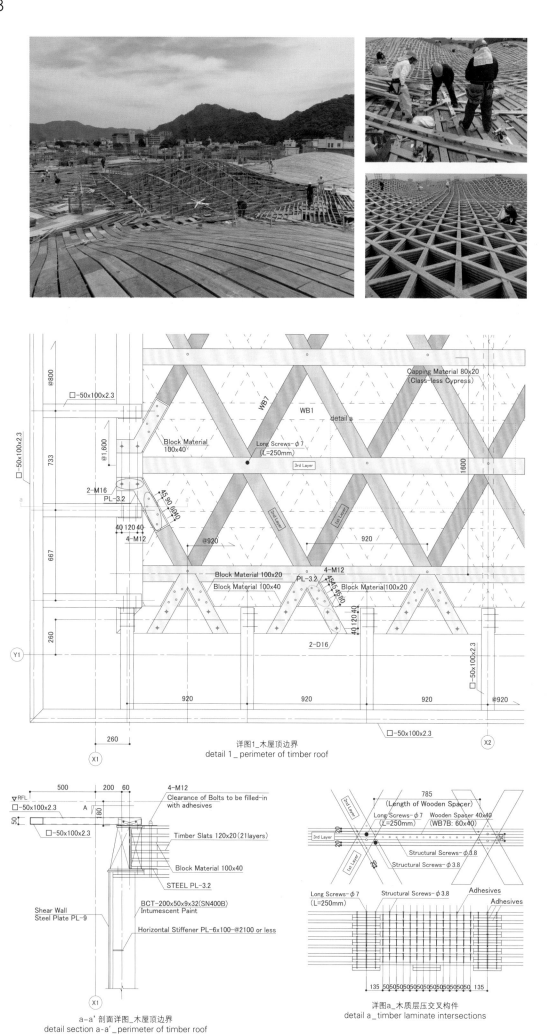

详图1_木屋顶边界
detail 1 _ perimeter of timber roof

a-a' 剖面详图_木屋顶边界
detail section a-a' _ perimeter of timber roof

详图a_木质层压交叉构件
detail a _ timber laminate intersections

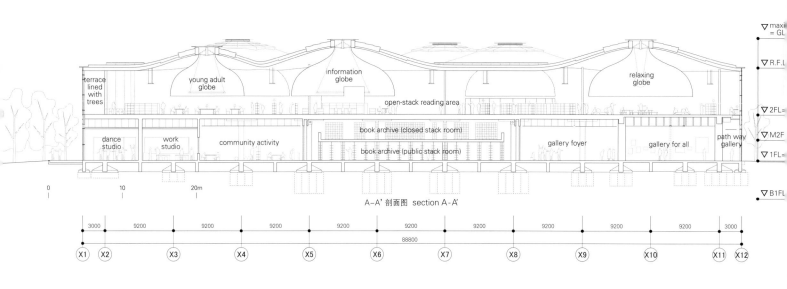

A-A' 剖面图 section A-A'

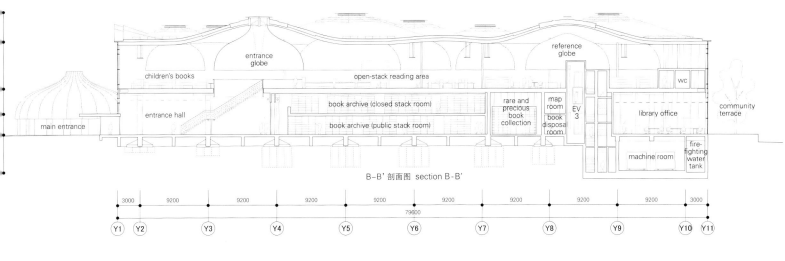

B-B' 剖面图 section B-B'

circular reading spaces as well as the reception counter. The "Globes" are made of polyester fabric and are shaped like an inversed funnel. Eleven "Globes" of 8m – 14m in diameter create gentle air movements and soft diffused light from above, providing a comfortable indoor environment.

The gently undulating wooden roof is made out of domestic-scale timber of 120mm x 20mm in section. The thin timber slats are stacked on top of each other, laminated in three directions. The function of "Globes" is optimized by the roof's undulation, as the roof rises where the "Globes" are located. The roof being embraced by the surrounding rows of mountains creates a grand and harmonious scene for the city. Plenty of trees were planted on the site, where there is a 240 metres long promenade of trees along the west side of the building. On the south, there is a plaza that is 45m wide. The each of the three terraces on the second floor has its own

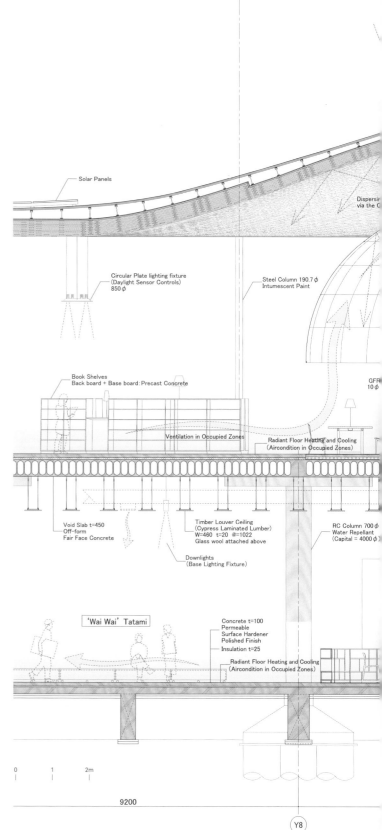

项目名称：Minna no Mori Gifu Media Cosmos
地点：Gifu, Japan
建筑师：Toyo Ito & Associates, Architects
设计团队：Toyo Ito, Takeo Higashi, Toyohiko Kobayashi, Yoshitaka Ihara, Toshimitsu Minami, Mitsuyo Yabuki, Ryo Chosokabe
结构工程师：Arup
施工单位：Toda·Dainippon, Ichikawa, Hinaya JV
机械工程师：ES Associates
电气工程师：Ohtaki E&M Consulting Office
其他顾问：Fujie Kazuko Atelier, Lighting Planners Associates, Nippon Design Center, Hara Design Institute, Nagata Acoustics, Ataka Fire Safety Design Office, Yoko Ando Design, Towa Prosperi, University of Tokyo Mikiko Ishikawa lab, Dainichi Consultant, Katsuhiko Hibino
客户：Gifu city
用地面积：14,848.34m²
有效楼层面积：15,444.23m²
材料：concrete(concrete sealer, densifier, chemical hardener), concrete block, glass, steel, wood, thermoplastic polyester triaxial woven fabric, special non-woven fabric, non-combustible interior materials, linoleum, carpet, vinyl chloride tile, diatomaceous earth
造价：JPY 12,500,000,000
设计时间：2011.2—2012.3
施工时间：2013.7—2015.2
摄影师：©Kai Nakamura (courtesy of the architect)

character, creating distinctively comfortable and relaxing areas. With the "Globes" channeling natural energy into the building, the utilization of underground water as a heat source to generate radiant floor heating and cooling, as well as the use of solar panels on the roof, all combined into a balanced mechanism. With all these features the building is expected to reduce the primary energy usage by at least 50%, compared to a building of the same size if it were in the 1990s.

Instead of clearly dividing or concealing spaces to reduce consumption of energy, we create a comfortable and energy-saving environment while pursuing connection with nature, a theme that has long existed in Japan, allowing people to feel alive and be nurtured within the space.

The users can enjoy the spaces as if there are in townscape and find new connections between people who use this space.

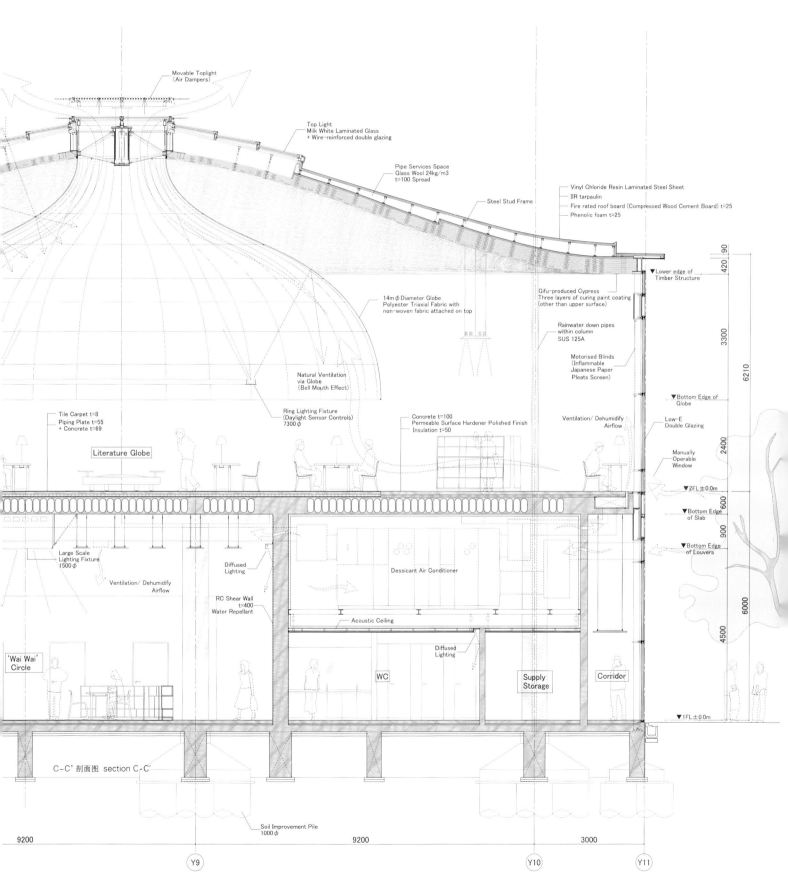

C-C' 剖面图 section C-C'

Saint-Just Saint-Rambert中央图书馆
Gautier + Conquet

该设施位于城镇的入口处，需要在多样化的城市肌理中寻找自己的定位。小镇内的房屋以个人房屋和小集群房屋为主，围绕一个大型广场而建，而这个大广场还在不断的发展中。因此，在这样的环境下，新媒体图书馆作为完善卫星城项目建设的一部分，也是45个村庄里的30多座图书馆中的最关键两项设施之一。

该图书馆周围将建有停车场、未来青年中心和南向广场，且一层包括一条通往原有的公园的人行道和大量绿化空间。

南面入口带有一个雨篷。该建筑是两层的大型礼堂，褶状屋顶让人联想到工厂的建筑风格。馆内最关键的设计是将整座建筑分隔的过厅，自然光通过这个过厅射入建筑。过厅打破了建筑的连续性，在整座建筑中心形成了一条贯穿建筑的通道，通道一层内有绿化天井，且一层完全被玻璃围合，并设有通往二层的三条通道。这里是建筑的中心，将图书馆的不同空间整合在一起。媒体图书馆仅位于上层，围绕一个单独的可调节的天台设置。一层是供大众和管理人员使用的，用于举办当地的社区活动，包括一处期刊阅读区，一间设有120个座位的展览厅和一座独立的礼堂。

二层空间非常宽敞、开放、明亮，采光主要依靠南北外立面上的大型窗户，再由另外两个立面的垂直洞口进行调节。褶状屋顶布满圆孔，由此引进顶部的光线，屋顶覆以轻木板，营造出一种温暖且平静的氛围。

空间规律地分布着彩色箱式构件。构件使用的建筑材料并不复杂，数量也受到限制，包括添加了石英的混凝土地面、木质隔声天花

东立面 east elevation

板，以及天井周围的玻璃隔墙。木材因其能效被选用来制作屋顶结构、木框架板和屋顶下的隔声板。这样的设计，不论冬夏，都能很好地控制室内温度，营造出舒适的环境。

Central Library in Saint-Just Saint-Rambert

At the entrance of the town, this facility needed to find its place in a varied urban fabric. With its individual houses, and small clusters of housing, the town is formed around a large square which continues to evolve. It is in this context that the new media library has been designed as part of the programme to complete the conurbation. It is one of the two key facilities in a network of around thirty libraries spread over 45 mainly rural villages.

The area around the building will hold a car park, the future youth centre, and a square to the south. The ground plan includes a pedestrian walkway to an existing park and will also welcome a lot of green spaces.

An awning marks the southern entrance. The building is a large two-storey hall, under the pleated roof that evokes

南立面 south elevation

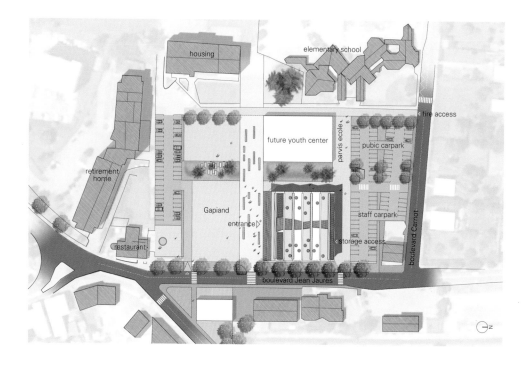

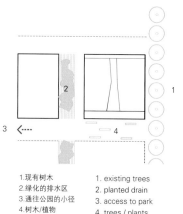

景观构成
landscape composition

1. 现有树木 — 1. existing trees
2. 绿化的排水区 — 2. planted drain
3. 通往公园的小径 — 3. access to park
4. 树木/植物 — 4. trees / plants

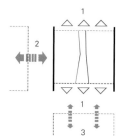

连接
connections

1. 公共立面 — 1. public facade
2. 与MJC直接连接的区域 — 2. direct connection with the MJC
3. 区域 — 3. place

可识别的轮廓
变化且分割的室内体量

identifiable silhouette
an interior volume both varying and dividing
第五立面 fifth facade

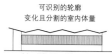

solar panels

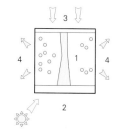

视野和照明
view and lighting

1. 屋顶光线和植被 — 1. roof light & vegetation
2. 前侧日光照明 — 2. frontal daylighting
3. 北侧日光照明 — 3. northern daylighting
4. 可见光区 — 4. visual lighting

公共电脑使用区
public use standalone operation

1. 礼堂 — 1. auditorium
2. 展厅 — 2. exhibition
3. 入口大厅 — 3. entrance hall
4. 卫生间 — 4. toilets

adult / children
multimedia / toilets

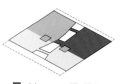

conference/exhibition / plantrooms
administration / press & magazine
interior garden / toilets

褶皱形覆盖屋顶
pleated cover

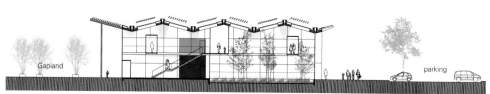

A–A' 剖面图 section A-A'

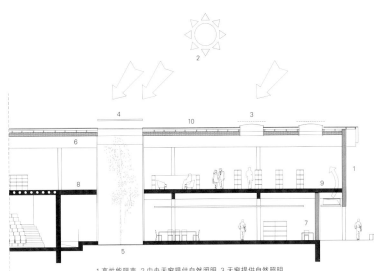

1.高性能隔声 2.中央天窗提供自然照明 3.天窗提供自然照明 4.遮光顶点 5.雨水回收区 6.木框架和木屋顶 7.木墙 8.混凝土板 9.双流通风管道 10.屋顶光伏板（可选）
1. high performance insulation 2. natural lighting by central skylight
3. natural lighting by skylights 4. sunscreen zenith 5. rainwater recovery
6. frame and wooden roof 7. wooden walls 8. concrete slab
9. ventilation double flux 10. PV (optional)

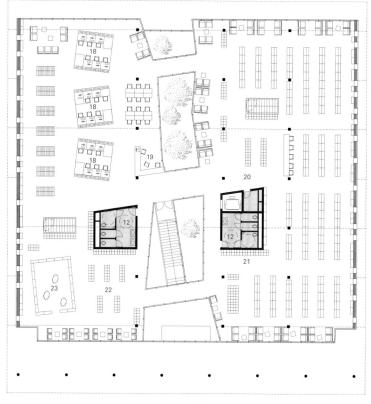

二层 second floor

1.北入口	1. north enterance
2.通道式储藏室	2. access storage
3.礼堂	3. auditorium
4.舞台	4. stage
5.展区	5. exhibition
6.自行车区	6. cycles
7.南入口	7. south enterance
8.媒体/杂志阅览区	8. press/magazines
9.接待处	9. reception
10.行政区	10. administration
11.工作间	11. workshop
12.卫生间	12. WC
13.员工休息室	13. staff break area
14.志愿者办公室	14. voluntary staff
15.储藏室	15. storage
16.服务间	16. server
17.发电室	17. plantroom
18.多媒体间	18. multimedia niche
19.借还书处	19. maindesk, returns
20.成人区	20. adult section
21.儿童区	21. children section
22.幼儿区	22. small children section
23.讲故事区	23. story telling section

一层 first floor

industrial architecture, and the key design element is a central rift separating the building to encourage natural light to enter the building. This rift deconstructs the uniformity of the building and creates a passage crossing the building. It contains a patio with greenery on the ground floor and is made of glass and holds three walkways on the second floor. It is the centrepiece that holds together the different spaces.

The media library has just one upper floor organised into a single and adaptable plateau. The ground floor is devoted to welcoming the public and administrative offices and actions to interest the local community: an area to read periodicals, an exhibition area with 120 seats and an independent auditorium.

On the upper floor, the volume is spacious, open and very light. The natural lighting mainly comes from the large windows on the south and north facades, tempered by the vertical openings of the two other facades. The pleated roof, dotted with circular oculi, creates the zenithal lighting. It is clad with panels of light wood, which give it a warm and calm ambience.

The space is punctuated by coloured box-type elements. The materials are unfussy and on a limited number: quartz concrete for the ground, wooden acoustic ceilings, and glass partitions around the patio.

Wood, chosen for its energy efficiency, is used in the roof structure, the wood-framed panels, and the acoustic panels beneath the roof. It gives the building a comfortable ambience due to the controlled temperatures in summer and in winter.

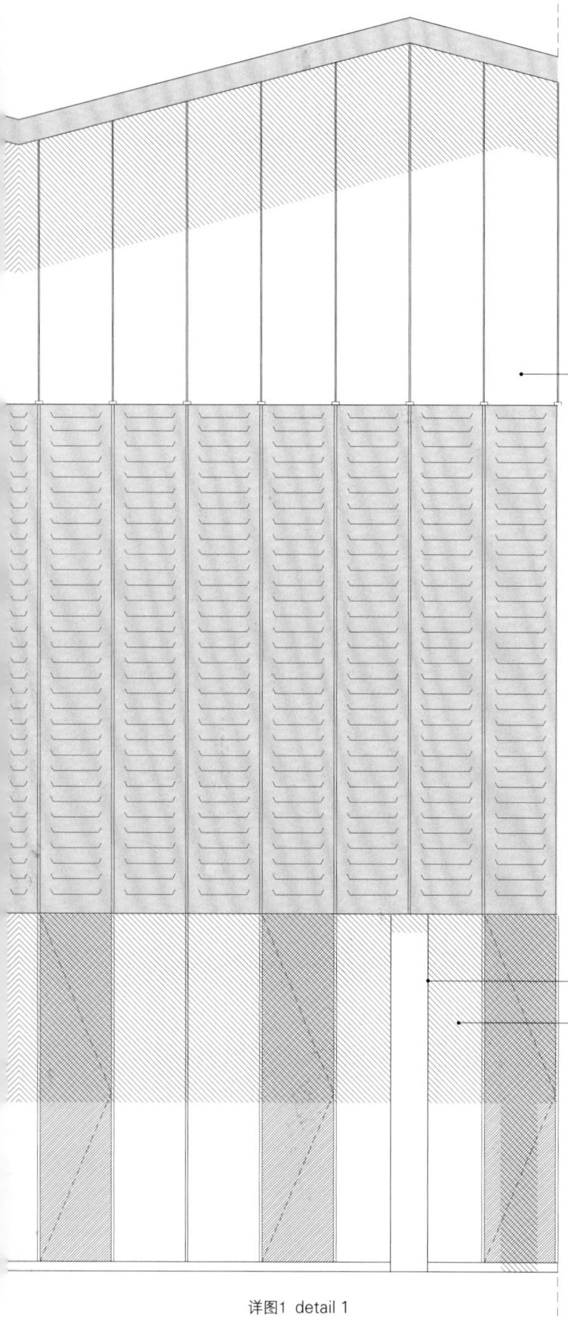
glazing opaque glass
concrete beam
glazing opaque glass

详图1 detail 1

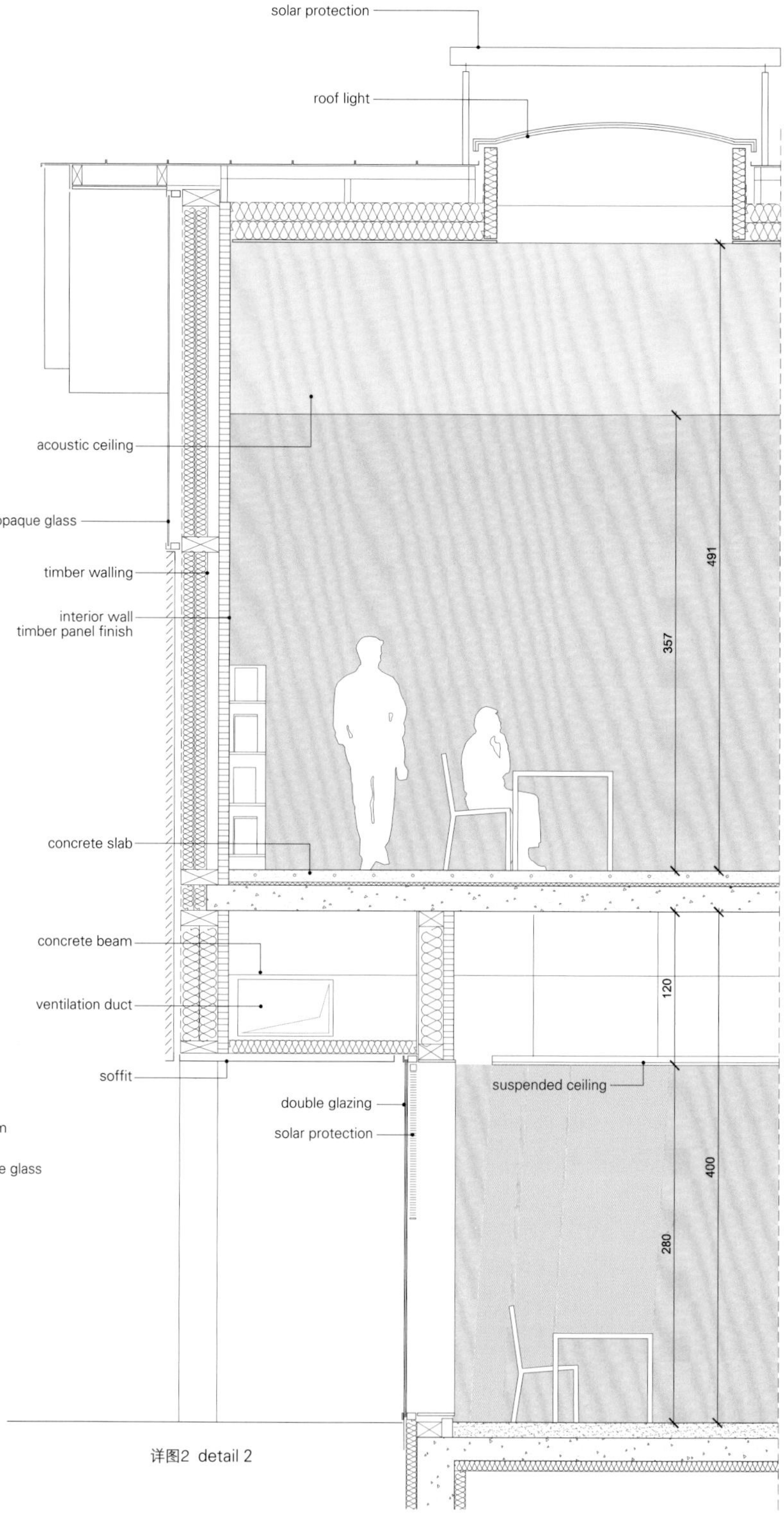
solar protection
roof light
acoustic ceiling
glazing opaque glass
timber walling
interior wall
timber panel finish
concrete slab
concrete beam
ventilation duct
soffit
double glazing
solar protection
suspended ceiling
491
357
120
400
280

详图2 detail 2

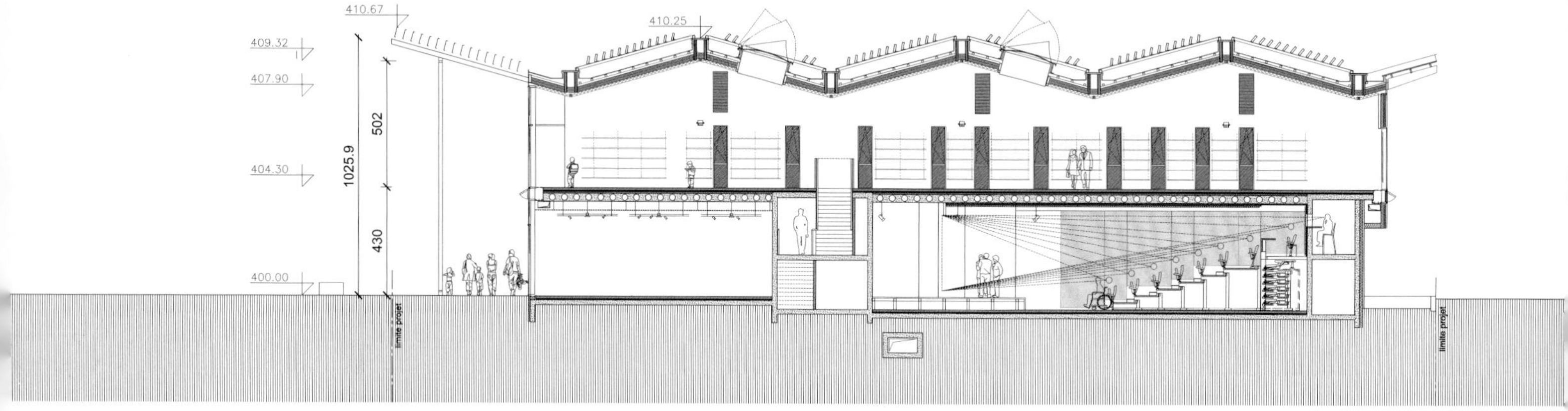
410.67
410.25
409.32
407.90
404.30
400.00
1025.9
502
430
limite projet
limite projet

B-B' 剖面图 section B-B'

项目名称：Central Library in Saint-Just Saint-Rambert / 地点：Place Gapiand à Saint Just Saint Rambert / 建筑师：Gautier+Conquet, architectes et paysagistes, Mandataire
用地面积：1,657m² / 总建筑面积：1,360m² / 有效楼层面积：2,272m²
设计时间：2010 / 施工时间：2013 / 竣工时间：2014 / 摄影师：©Michel Denancé(courtesy of the architect)

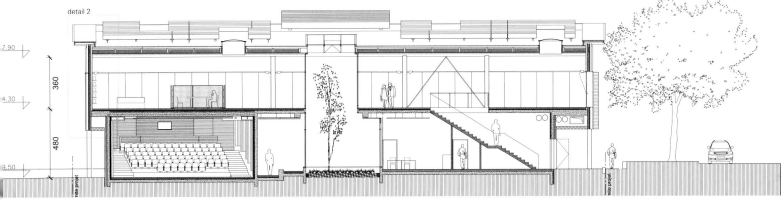

C-C' 剖面图 section C-C'

新型社区图书馆 New Community Library

Constitución公共图书馆
Sebastián Irarrázaval Arquitectos

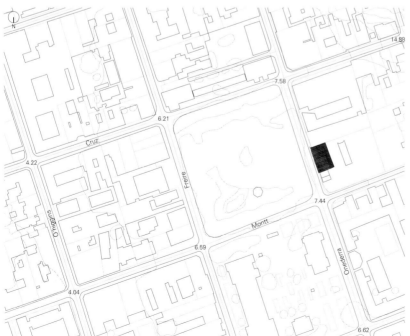

Constitución公立图书馆是在2010年Constitución市遭受8.8级地震和海啸（损毁了这座城镇）后所进行的灾后城市重建的一个项目，由公共机构和私人募捐共同发起。这里是一个小型定居地，坐落在智利最大的木材产地的中心区。这样为寻找高质量的木材提供最优越的条件，同时这里还有很多天赋异禀的木匠，能够设计出精美的木建筑。

此建筑项目得益于三个主要的决定。第一，为了让人们可以在图书馆中俯瞰场地前面的人民广场上的千年古树，对比街道水平面，图书馆整体提高1.6m。第二，为了过滤和平衡光线，建三个带有网状屋顶的正厅，分布在三处主要功能区。第三，为了与大楼的公共特征相呼应，立面布局为三个大型的玻璃书柜，不仅用来安放新书籍，还安置了长椅和雨篷，供访客利用。

从该建筑的建设方面来说，建筑整体主要是由木材构成，只有防火墙表面为浇灌的裸露混凝土。整体的结构是预先打造的，材质为层压松木构件，为了赋予内部空间的一定的节奏，承重测算和施工过程清晰易懂，木横梁和支柱都尽可能地暴露在外面。木材都涂白色透明油漆，既提高了内部所需的亮度，又使结构和场地内打造的家具和协搭配。建筑结构中其他可见的颜色则采用与广场上的树木和叶子颜色相近的颜色。从这个意义来说，整座图书馆就是一处产生共鸣的空间。

Constitución Public Library

Constitución Public Library is part of a public-private initiative taken to rebuild the city of Constitución after the 8.8 degrees earthquake and tsunami that devastated the town during the year 2010. It is a small settlement situated in the very core of one of the biggest clusters of wood production in Chile. This identity also creates the best conditions to find not only high quality wood materials but also extremely well gifted carpenters to carefully craft a wooden building.

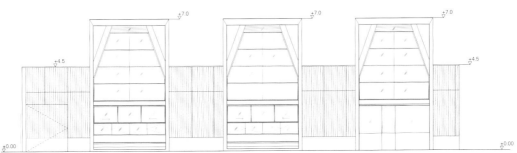

东北立面 north-east elevation

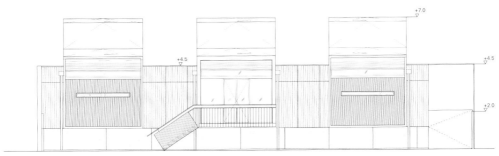

西南立面 south-west elevation

项目名称：Public Library of Constitución
地点：Constitución, Chile
建筑师：Sebastian Irarrázaval
合作：Macarena Burdiles, Carlos Pesquera, Alicia Arguelles, Sebastián Mancera
技术顾问：Joel Barrera
发起人：Fundación la Fuente, Banco Itau and Arauco
业主：Municipality of Constitución
结构工程师：Cargaz
承包商：PROESSA
用地面积：400m² / 有效楼层面积：350m²
设计时间：2011 / 竣工时间：2015
摄影师：©Felipe Díaz Contardo

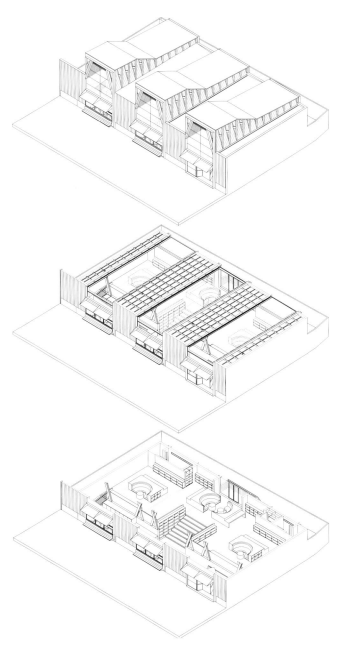

With regards to the formalization of the project; it is the result of three main decisions. Firstly: In order to overlook the millenary trees of the civic square in front of the site; the library is risen 1,6 meters over street level. Secondly: With the purpose to filter and balance the light; the 3 main areas of the programme (children, young and adult readers) were covered with 3 reticulated wood naves and Thirdly: In consideration to communicate the public character of the edifice the facade was organized with 3 monumental glass cases that not only invite new book arrivals but also – with its benches and canopies – offer shelter to the passersby.

Regarding the construction of the building; it is made almost entirely of wood and only the firewalls are done with exposed poured concrete. The structure is prefabricated and is made out of laminated pine. In order to give rhythm to the interior space and to make the loads and the construction process understandable, the wooden beams and pillars are kept as visible as possible. Coating the wood with transparent white varnish enhances the required luminosity of the spaces and also creates homogeneity between the structure and the on site built furniture. The other colors that can be seen in fabrics were chosen to mimic the colours of the trees and leaves of the square. In this sense the library can be seen as a resonance box.

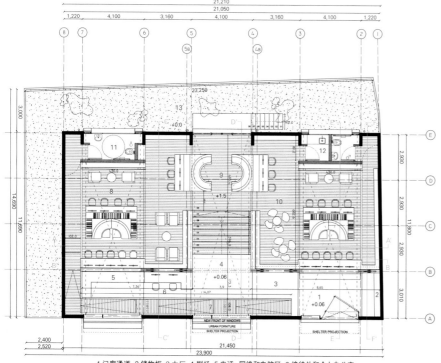

1.门廊通道 2.储物柜 3.大厅 4.剧场 5.电话、网络和电脑区 6.接待处和个人办公室
7.斜坡 8.青年区 9.报纸和杂志阅览区 10.儿童区 11.残疾人卫生间 12.儿童和储藏室 13.天井
1. Zaguan access 2. locker storage 3. hall 4. amphitheater 5. telephones, nets, and computer area
6. reception and personal office 7. ramp 8. young area 9. daily newspaper and magazine reading
10. children area 11. toilet with disabilities 12. kitchen and storage 13. patio

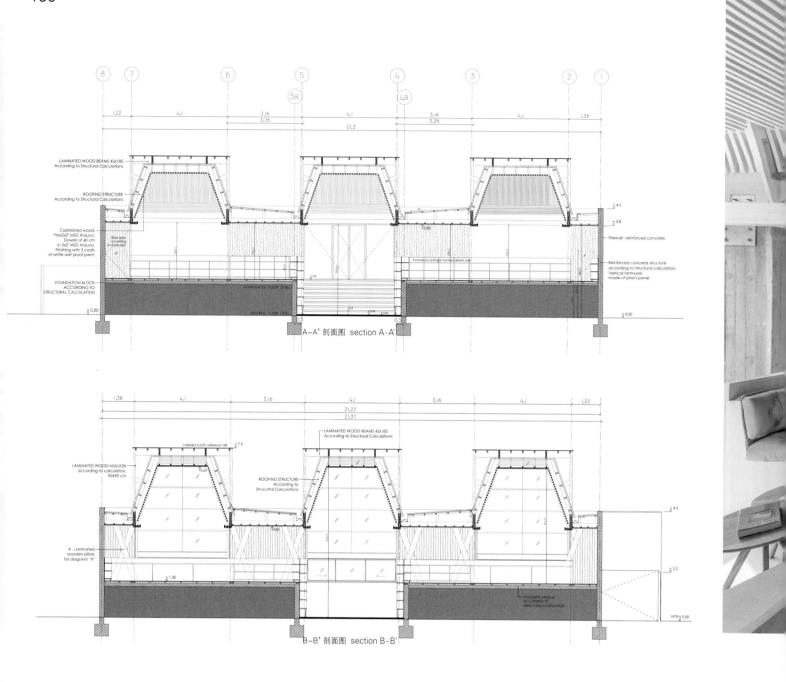

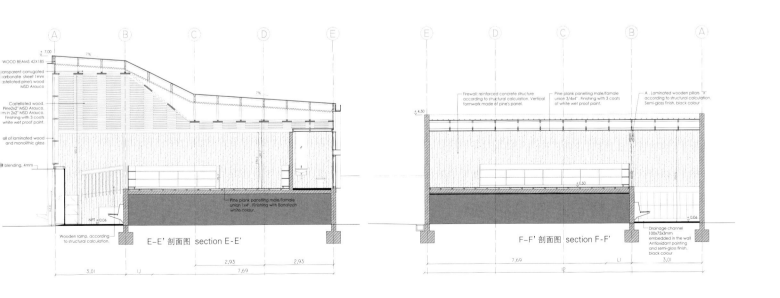

E-E' 剖面图 section E-E'

F-F' 剖面图 section F-F'

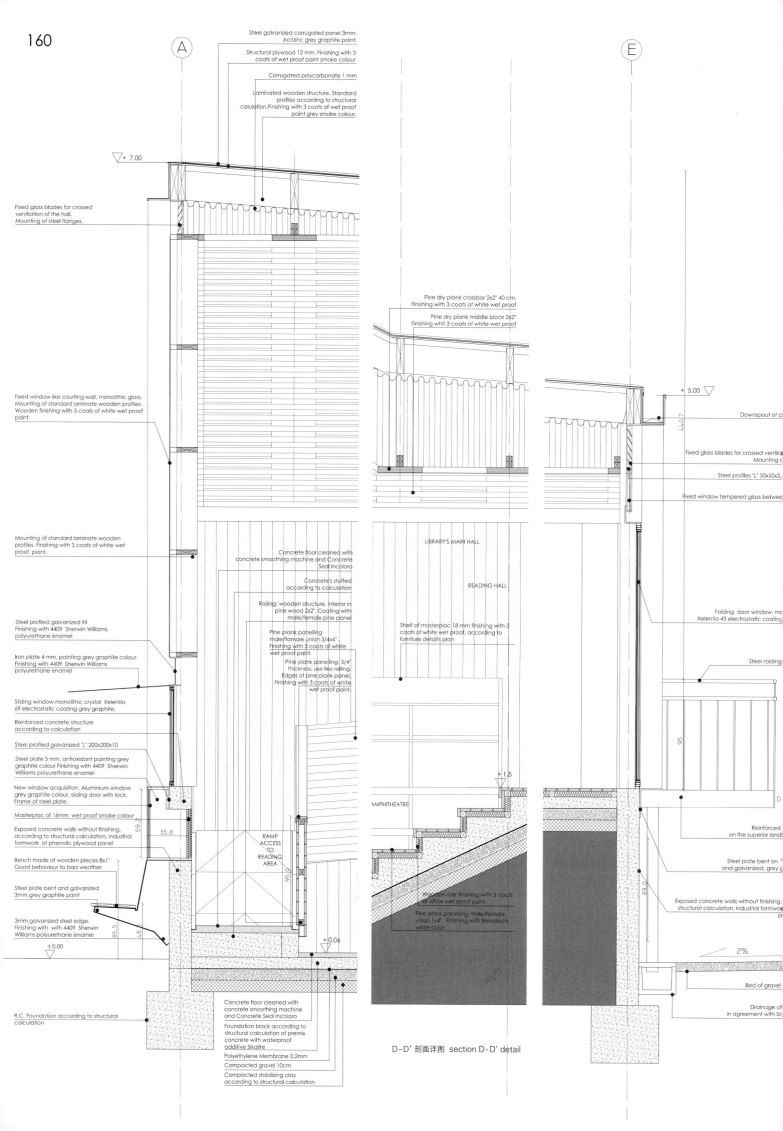

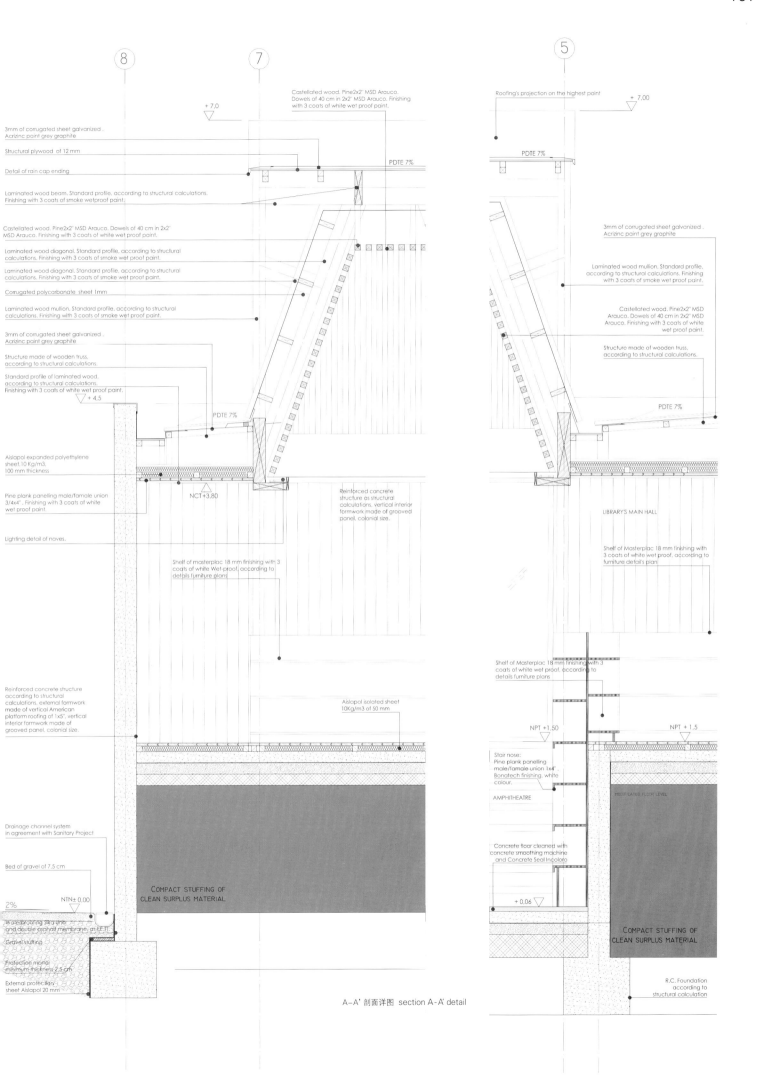

A-A' 剖面详图 section A-A' detail

新型社区图书馆 New Community Library

海岸图书馆
Vector Architects

这座海岸图书馆距北京约三小时车程，坐落于渤海湾的一个度假村内。虽然北京在经济和城市发展上取得了巨大的进步，但是许多人却都意识到北京的居住环境水平在下降。于是，为了迎合这一需求，这座度假村为大众创造了更贴近大自然的生活环境。在这里，有一系列的文化和休闲娱乐设施，海岸图书馆就是其中一项。

设计的核心理念就是探索和谐共存关系，将空间界限、人类活动、光照调节、空气流动和海洋景色有机结合。图书馆东面朝向大海，在春、夏、秋三季，它不仅在西侧接待社区居民，还会对大众开放。

图书馆中有各个功能区，有阅读区、静思厅、活动室、酒吧和休息区。不同的区域，根据其不同的空间，有不同的设计，不同的通风模式和灯光控制，以及不同的欣赏海景的角度。

随着季节的变化，昼夜的更替，海洋在不断地发生变化，像是大自然的演出。一层，图书馆面朝大海的一面被一系列玻璃墙围合。天气好的时候，这面玻璃墙就会被打开，让内部空间和外面的大海直接连接。而在这些可活动的玻璃墙上方，是一整块水平的观赏窗，横穿整个图书馆，这是观赏海景的主屏幕。为了避免结构组件的断裂，位于观赏窗的上部屋顶承重都由钢桁架支撑的。在钢桁架结构两侧，我们用人工玻璃砖填充墙面，这样的玻璃砖设计会柔化坚硬的钢桁架。而且，这种半透明的材料对光线很敏感，它会根据天气的不同转换自然光线和人工光线，使建筑的氛围变得特别柔和。

静思厅在阅读区的旁边，与明亮开放的阅读区不同，静思厅光线比较昏暗，明暗对比强烈，提供一个封闭而私密的空间。只有两条30cm宽的通气口，一个在房间的东面，一个在房间的西面；一个是水平，一个是垂直；一个用来捕捉日出的晨曦，另一个用来吸收落日的余晖。大厅的屋顶是大幅度的曲面设计，并使天花板向下延伸。屋顶上还设置了一个低露台。在这里，尽管欣赏不到海景，人们还是可以听到海浪的声音。

活动室，由于其不同的用途，可能会产生噪音，所以与阅读区之间隔了一个室外的平台，形成了一个独立的空间。屋顶上朝东的采光井和朝西的通风窗能够从不同的方向获取光线，冷暖光相互重叠着，给整个空间着色。

Seashore Library

About three hours drive from Beijing, the library is located inside a vacation compound along Bohai Bay. While Beijing has been experiencing the massive growth in economy and city development, many have pointed out the issue of drop in living environment. The vacation compound aims to create a quality-living area closer to the nature. There are a series of cultural and leisure facilities within the compound, and Seashore Library is one of them.

The design key point is focused on exploring the co-existing relationship of the space boundary, the movement of human body, the shifting light ambiance, the air ventilating through and the ocean view. The library faces the ocean to its east. During seasons of spring, summer and fall, it not only serves the community residents at west, but opens to the public as well.

The library houses a reading area, a meditation space, activity room, a drinking bar and a resting area. According to each space, we establish distinctive relationship between space and the ocean, and define how light and wind enters into each room.

Ocean, an ever-changing character continues to alter from season to season, morning to night. It is like a drama of nature. Toward the sea, the building is enclosed by a series of glass walls at ground level. When the weather is nice, the walls are open to the sea directly connecting inside and outside. On

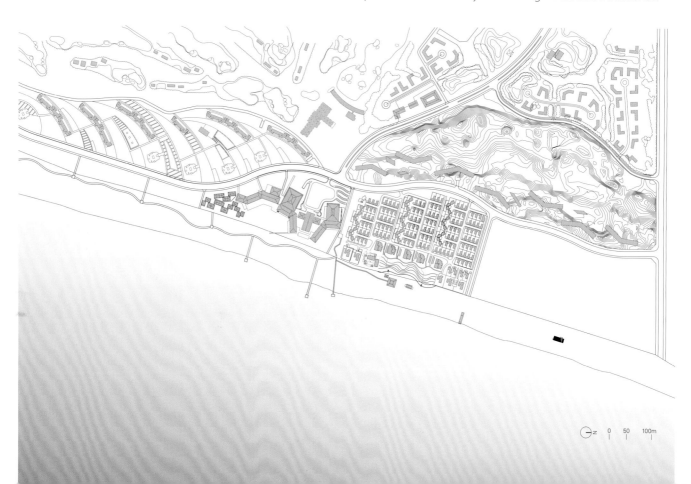

东立面 east elevation

北立面 north elevation

西立面 west elevation

南立面 south elevation

top of these pivot walls is a horizontal view window that goes across the library; it is the main frame of sea view. To avoid interruption from any structural component, all the roof loads are carried by the steel trusses running above the view window. On both sides of the steel trusses, we infill hand-crafted glass block masonries into wall. The wall softens the hardness of steel trusses. Furthermore, the translucency of such material is sensitive to light. It transforms both natural and artificial light to inside and outside throughout different of the day, smoothly changing the ambience of the building.

The meditation space sits aside from the reading area. In contrast to the evenly bright open and public reading area, the meditation space is rather dim, with sharp light and shadows, enclosed and private. There are two slim openings, 30cm

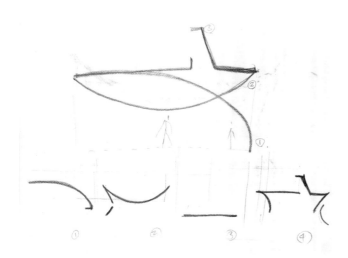

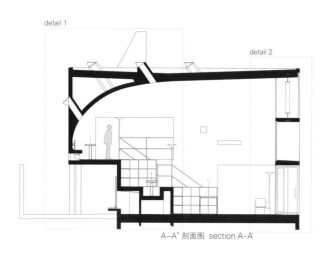
A-A' 剖面图 section A-A'

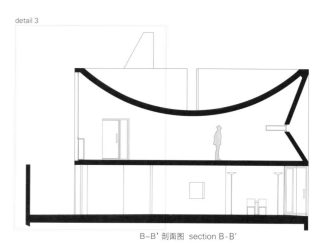
B-B' 剖面图 section B-B'

C-C' 剖面图 section C-C'

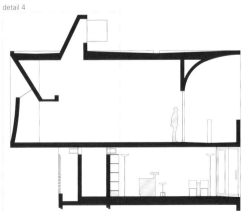

detail 4

D-D' 剖面图 section D-D'

wide, on east and west side of the room. One is horizontal and one in vertical; one captures the light of sunrise and the other grasps the sunset illumination. In this room, a drastic curved roof pushes the ceiling down. Above this curve, it creates a low terrace on the roof top. At this area, people hear the sound of ocean, though the vision is out of reach.
The activity room is a fairly isolated space. Due to potential sound, (it is separated) from the reading area with an outdoor platform in between. The light well on the roof facing the east and the clear story at west collect light throughout the day from different directions. Warm and cold light overlap and tint the space simultaneously.

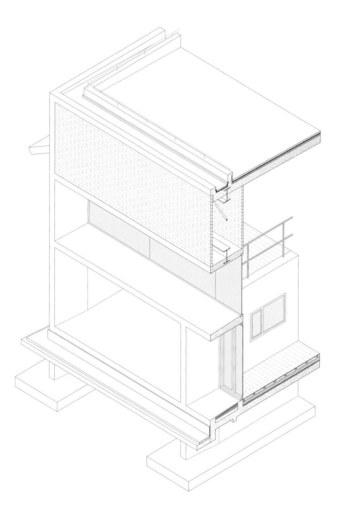

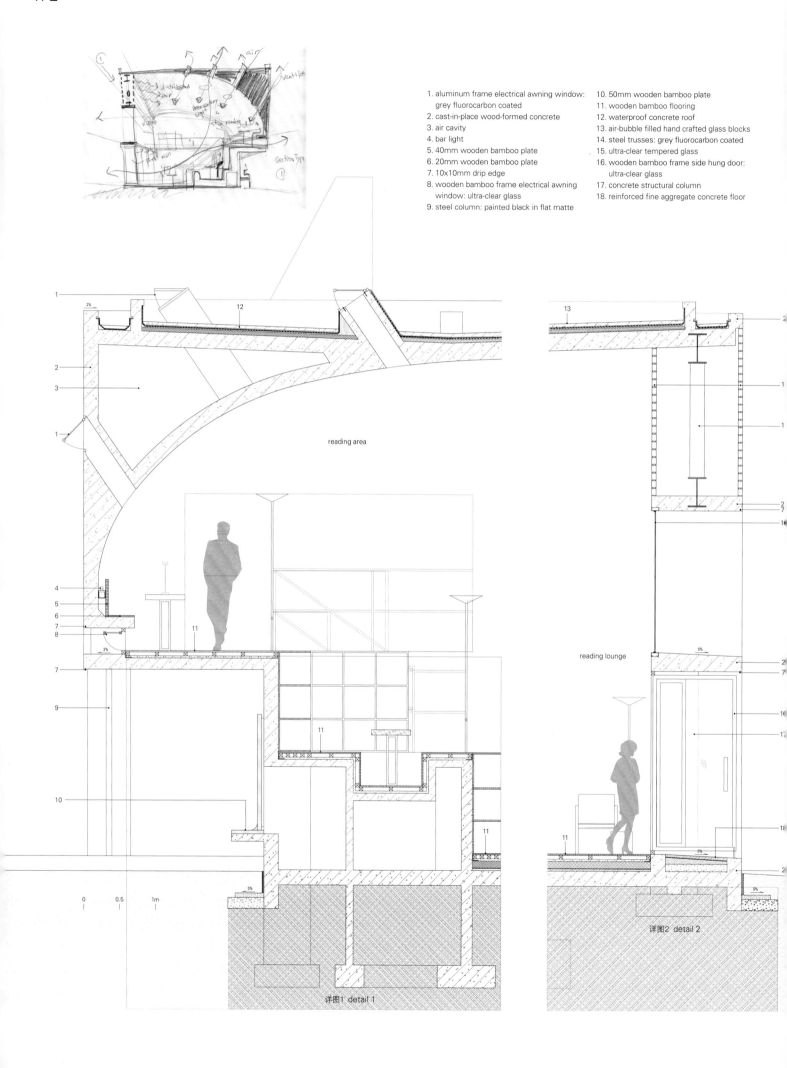

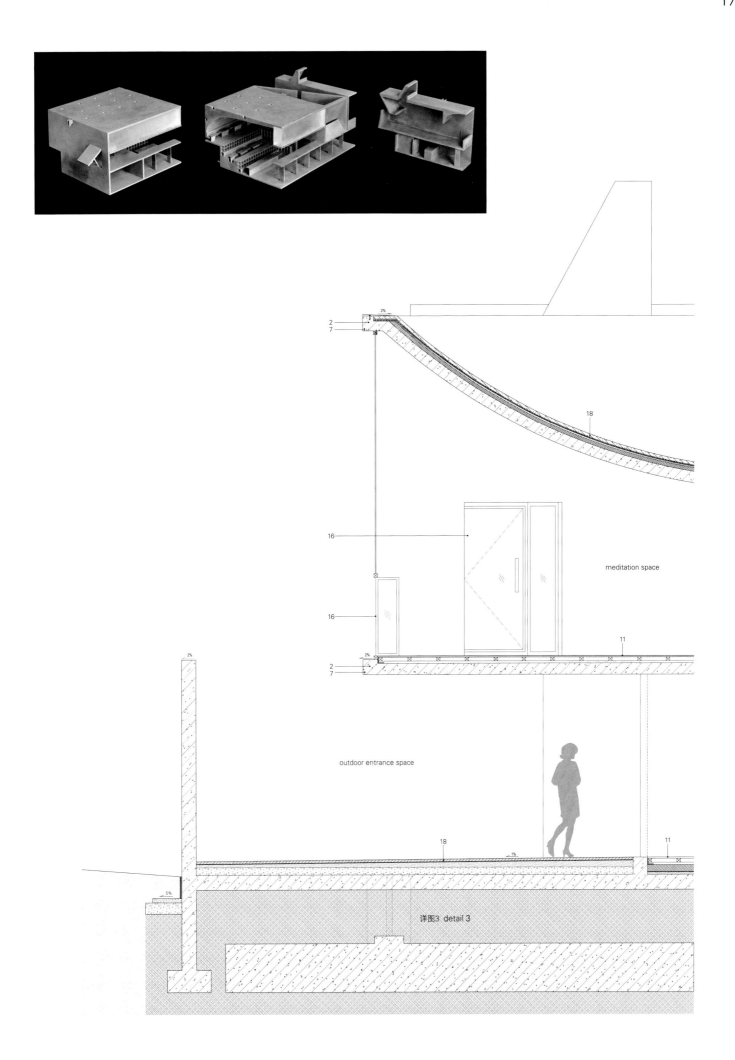

详图3 detail 3

项目名称：Seashore Library / 地点：Beidaihe New District, China
建筑师：Vector Architects / 主要建筑师：Gong Dong
项目建筑师：Chen Liang / 场地建筑师：Yifan Zhang, Dongping Sun / 设计团队：Zhiyong Liu, Hsi Chao Chen, Hsi Mei Hsieh
结构&MEP工程师：Beijing Yanhuang International Architecture & Engineering Co.,Ltd.
结构顾问：Lixin Ji, Zhongyu Liu / 客户：Beijing Rocfly Investment (Group) CO., LTD
总建筑面积：450m²
材料：concrete, laminated bamboo slate, glass block masonry
设计时间：2014.2~7 / 施工时间：2014.7~2015.4
摄影师：
Courtesy of the architect-p.177, p.179[top], ©He Bin (courtesy of the architect)-p.170~171, p.174[bottom]
©Su Shengliang (courtesy of the architect)-p.162~163, p.166~167, p.168~169, p.175,
©Xia Zhi (courtesy of the architect)-p.165, p.174[top], p.176, p.178~179, p.179[bottom], p.180~181

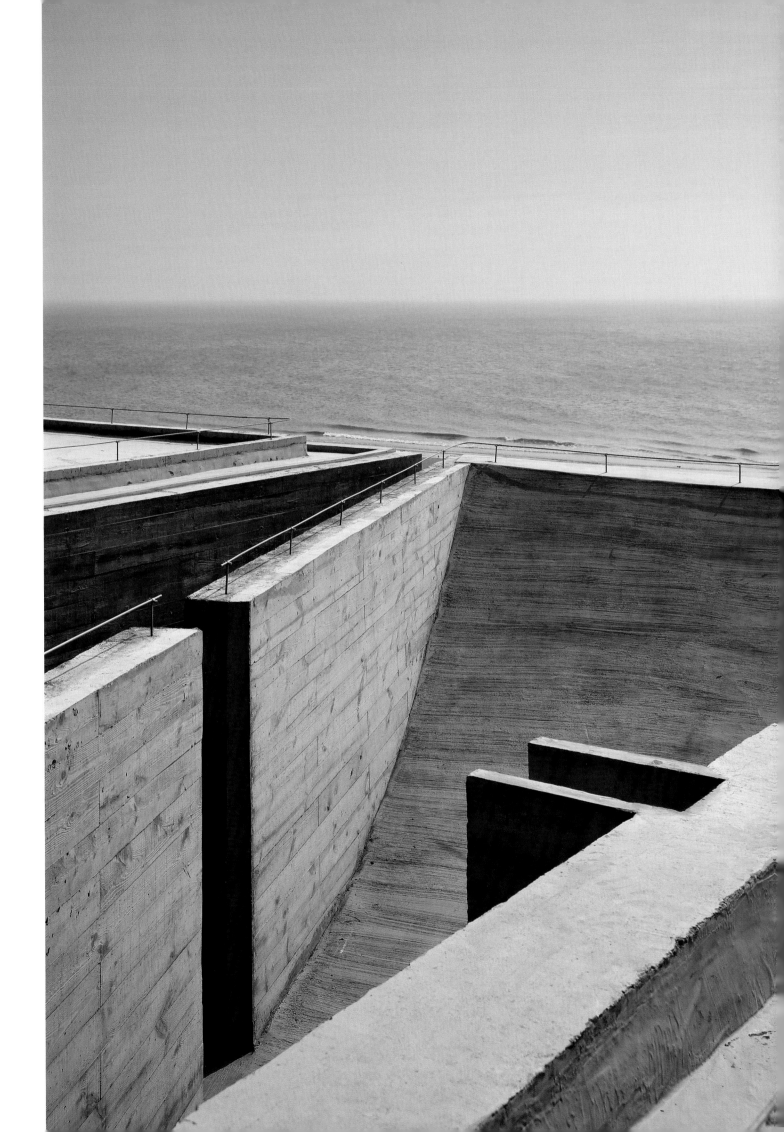

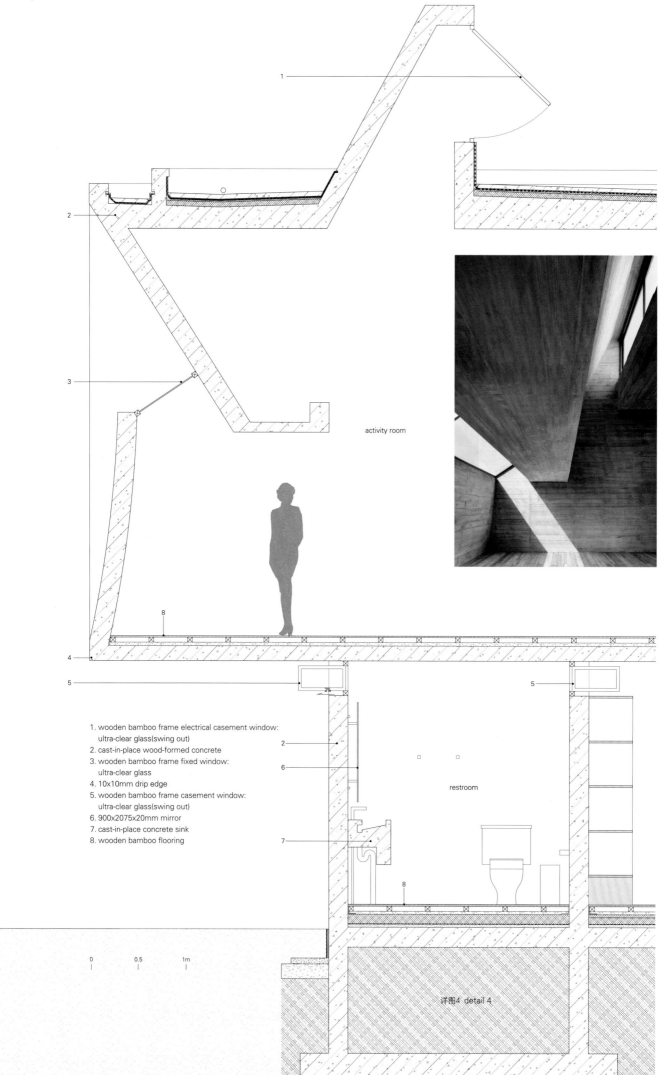

activity room

1. wooden bamboo frame electrical casement window: ultra-clear glass(swing out)
2. cast-in-place wood-formed concrete
3. wooden bamboo frame fixed window: ultra-clear glass
4. 10x10mm drip edge
5. wooden bamboo frame casement window: ultra-clear glass(swing out)
6. 900x2075x20mm mirror
7. cast-in-place concrete sink
8. wooden bamboo flooring

restroom

详图4 detail 4

1.接待处 2.图书展示区 3.阅读走廊 4.休息区 5.酒吧 6.卫生间 7.储藏室 8.办公室 9.室外区域
1. reception 2. book display area 3. reading lounge 4. resting area 5. bar 6. toilet 7. storage 8. office 9. outdoor area
一层 first floor

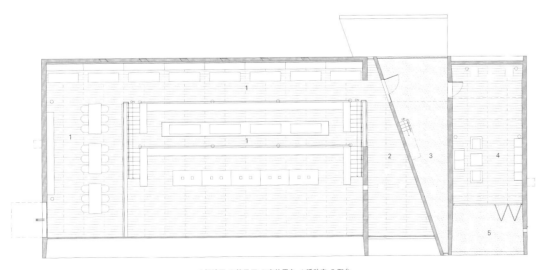

1.阅读区 2.静思厅 3.室外平台 4.活动室 5.阳台
1. reading area 2. meditation space 3. outdoor platform 4. activity room 5. balcony
二层 second floor

>>10

Wespi de Meuron Romeo Architects
Was founded by Markus Wespi[left], Jérôme de Meuron[middle] and Luca Romeo[right] in 2012. Markus Wespi was born in 1957 in St. Gallen, Switzerland. He had learned architecture for himself. Has been working with Jérôme de Meuron since 1998. Jérôme de Meuron was born in 1971 in Münsingen, Germany and studied at the Technical School of Burgdorf from 1993 to 1996. He experienced architectural practical while staying in Ghana, Africa from 1996 to 1997. Luca Romeo was born in 1984 in Locarno, Switzerland and studied at the Technical School of Lugano-Trevano from 2003 to 2006. He has been working with Markus Wespi and Jérôme de Meuron since 2011.

>>100

Christ & Gantenbein
Was established in 1998 by Swiss architects, Emanuel Christ and Christoph Gantenbein. International reputation of Christ & Gantenbein was consolidated by renovation and extension of the Swiss National Museum in Zürich, Kunstmuseum (Museum of Fine Arts) in Basel and Wallraf-Richartz-Museum in Köln. Major part of the practice focuses on museums and their thematic research on dealing with both the old and the new has remained in focus. Besides their practical activities, they have always been lecturing at various universities including ETH Studio Basel, HGK in Basel, Mendrisio Academy of Architecture, the Oslo School of Architecture and Design, Harvard GSD and their alma mater, ETH Zürich.

>>136

Gautier + Conquet
Chief Executive, Dominique Gautier(1960) and General Manager, Stéphane Conquet(1968) are French Architects DPLG qualified in public spaces, urban furniture, transport infrastructures, public facilities and housing. Dominique Gautier is a graduate of the National Superior School of Architecture at Lyon. Associated with Bruno Dumetier, AABD Architects in 1999. Is a member of the AMO Association(Architecture et Maîtres d'Ouvrage). Stéphane Conquet is a graduate of the National Superior School of Architecture of Paris la Villette. Joined the Association with Bruno Dumetier and Dominique Gautier, AABD Architects in 2003. Created the Parisian Studio in 2004. AABD Architects became Gautier + Conquet & Associés in 2008.

©Markus Jans

>>162

Vector Architects

Gong Dong graduated from Tsinghua University, Department of Architecture in 1999 and received master's degree in architecture at the University of Illinois in 2001. Prior to establishing his own practice he worked for Soloman Cordwell Buenz & Associates in Chicago, then at Richard Meier & Partners and Steven Holl Architects in New York. In 2008, he founded Vector Architects and has become one of the most active young architects in China. Is currently teaching at Tsinghua University, School of Architecture.

>>148

Sebastián Irarrázaval Arquitectos

Sebastián Irarrázaval was born in 1967. Studied at the Catholic University of Chile and the Architectural Association in London. In 1993, he set up his own practice in Santiago and has been teaching at the Catholic University of Chile since 1994. Has also taught at the MIT, Boston and the IUAV, Venice. Received the AOA(Architecture Offices Association) award for the most outstanding young architects and was awarded in the XVI Architecture Biennial in Chile. Was also awarded twice at the Wave International Workshop held at IUAV, Venice in 2014 and 2015. This year, he was awarded honor at the Wood Design and Buidling Award for the Constitución Public Library.

>>54

Gion A. Caminada

Was born 1957 in Vrin, Switzerland. Following a carpenter apprenticeship in Vrin, he attended the Kunstgewerbeschule in Zürich and then went on to complete postgraduate studies in architecture at the ETH Zürich. Since 1986, Caminada has been running his own architectural practice in Vrin. In 1999, he was named as an assistant professor at ETH Zürich, and since 2008, he has been an associate professor for architecture.

>>86

mlzd

Was established in Bienne, Switzerland in 1997. Is the workplace of a versatile team of architects, who won more than 30 first prizes in international competitions and have more than 40 completed building projects to their name. 30 employees from different cultures, ages and experience promote a multifaceted office culture. Their internal discussions give rise to vastly varied projects. Most important projects include the renovation of the presidential anterooms to the United Nations General Assembly Hall in New York(2004) as well as extensions to the Historic Museum in Bern(2009) and the Local-Heritage Museum in Rapperswil(2011).

>>120
Toyo Ito & Associates, Architects
Toyo Ito graduated from the University of Tokyo, Department of Architecture in 1965. In 1971, he established his own office Urban Robot which was renamed Toyo Ito & Associates, Architects in 1979. Main works are Sendai Mediatheque, Tama Art University Library, Torres Porta Fira, Toyo Ito Museum of Architecture, Imabari, National Taiwan University, College of Social Sciences, Minna no Mori Gifu Media Cosmos, etc. He was awarded Golden Lion for Lifetime Achievement from the 8th International Architecture Exhibition 'NEXT' at the Venice Biennale, Royal Gold Medal from the Royal Institute of British Architects, Golden Lion for Best National Participation for the Japan Pavilion from the 13th International Architecture Exhibition at the Venice Biennale and the Pritzker Architecture Prize.

>>40
Nickisch Walder
Selina Walder[left] made her diploma at the Mendrisio Academy of Architecture with Valerio Olgiati in 2004. From 2004 to 2006, she taught as an assistant at his chair for architectural design. Since 2005, she has been working with Georg Nickisch in Flims. Georg Nickisch[right] studied at the Prince of Wales Institute of Architecture, The University of Bath and made his degree in architecture at the Mendrisio Academy of Architecture with Peter Zumthor in 2005. From 2007 to 2013 he taught as an assistant at the chair of Jonathan Sergison (Sergison Bates Architects). Currently he teaches as a guest professor at the HEAD (Geneva University of Art and Design) Geneva.

>>62
Pascal Flammer
Was born in Fribourg, Switzerland in 1973. Studied architecture at the ETH Zürich, EPF Lausanne and TU Delft before receiving a Master of Science in Architecture from the ETH Zürich in 2001. Pascal gained his experience through his time at Valerio Olgiati and opened his practice in 2005. Became a Member of SIA (Swiss Engineers and Architects Association) in 2005. Has taught at the Mendrisio Academy of Architecture, the GSD Harvard and at the Sandberg Institution in Amsterdam. Is currently teaching at the ETH Zürich. Received various awards including Philippe Rotthier European Prize for Architecture 2014 and 2013 Architecture Award Kanton Solothurn, Switzerland.

Anna Roos
Studied architecture at the UCT(University of Cape Town, South Africa) and holds a postgraduate degree from the Bartlett School of Architecture, UCL, London. Moving to Switzerland in 2000, worked as an architect, designing buildings in Switzerland, South Africa, Australia, Scotland. Has been working as a freelance architecture journalist since 2007 and writes for A10, Ensuite Kultur Magazin, Monocle, and Swisspearl.

Tom Van Malderen
His activities stretch from the traditional architectural practice to the field of architectural theory which he explores through writing, installations and lectures. After obtaining a master in Architecture at LUCA, Brussels(1997) he worked for Atelier Lucien Kroll in Belgium and in different positions at architecture project, both in the UK and Malta. Lectured at the University of Aix-en-Province in France and the Canterbury University College of Creative Arts in the UK. Contributes to several magazines and publications, and sits on the board of the NGO inemastik for the promotion of short film.

>>30
Alp' Architecture
Is based and active in Bagnes since 2010 and in Lausanne since 2014. Is run by Laurent Berset[right], Sacha Martin[middle] and Romain Pellissier[left]. The office proposes a complete take over of projects of all scales and stands behind an architecture bound to the social and cultural context, in the same time contemporary, innovative and respectful of the location. Laurent and Sacha received a Master of Architecture at the Federal Institute of Technology Lausanne. Romain received a Bachelor of Architecture at the HES-SO(University of Applied Sciences and Arts at Western Switzerland). He is a member of the Architectural Commission of the municipality of Bagnes.

>>76
Andreas Fuhrimann Gabrielle Hächler Architekten
Andreas Fuhrimann[second] graduated from ETH Zürich in 1985 and worked as design and planning architect at the Marbach + Rüegg. Gabrielle Hächler[first] also graduated from from ETH Zürich in 1988 and worked as an assistant lectureship at her school. They have been co-operating the architectural office since 1995 and were admitted to the Association of Swiss Architects(BSA) in 2005. Carlo Fumarola[third] has been working at Fuhrimann Hächler Architects since 2005 after receiving Architectural degree at the ETH Zürich. Gilbert Isermann[fourth] also graduated from the ETH Zürich in 2004 and has been working at Fuhrimann Hächler Architects since 2007. Carlo Fumarola and Gilbert Isermann are Partners at Fuhrimann Hächler Architects since 2012.

© 2016大连理工大学出版社

图书在版编目(CIP)数据

探索瑞士建筑的异曲同工之妙：汉英对照 /（瑞士）安娜·鲁斯等编；安雪花等译. — 大连：大连理工大学出版社，2016.12
 ISBN 978-7-5685-0602-1

Ⅰ. ①探… Ⅱ. ①安… ②安… Ⅲ. ①建筑艺术－瑞士－图集 Ⅳ. ①TU-865.22

中国版本图书馆CIP数据核字(2016)第291297号

出版发行：大连理工大学出版社
　　　　（地址：大连市软件园路80号　邮编：116023）
印　　刷：上海锦良印刷厂
幅面尺寸：225mm×300mm
印　　张：11.75
出版时间：2016年12月第1版
印刷时间：2016年12月第1次印刷
出 版 人：金英伟
统　　筹：房　磊
责任编辑：许建宁
封面设计：王志峰
责任校对：高　文
书　　号：978-7-5685-0602-1
定　　价：228.00元

发　行：0411-84708842
传　真：0411-84701466
E-mail：12282980@qq.com
URL：http://www.dutp.cn

版权所有·侵权必究

C3 建筑立场系列丛书 01:
墙体设计
ISBN: 978-7-5611-6353-5
定价: 150.00 元

C3 建筑立场系列丛书 02:
新公共空间与私人住宅
ISBN: 978-7-5611-6354-2
定价: 150.00 元

C3 建筑立场系列丛书 03:
住宅设计
ISBN: 978-7-5611-6352-8
定价: 150.00 元

C3 建筑立场系列丛书 04:
老年住宅
ISBN: 978-7-5611-6569-0
定价: 150.00 元

C3 建筑立场系列丛书 05:
小型建筑
ISBN: 978-7-5611-6579-9
定价: 150.00 元

C3 建筑立场系列丛书 06:
文博建筑
ISBN: 978-7-5611-6568-3
定价: 150.00 元

C3 建筑立场系列丛书 07:
流动的世界:日本住宅空间设计
ISBN: 978-7-5611-6621-5
定价: 200.00 元

C3 建筑立场系列丛书 08:
创意运动设施
ISBN: 978-7-5611-6636-9
定价: 180.00 元

C3 建筑立场系列丛书 09:
墙体与外立面
ISBN: 978-7-5611-6641-3
定价: 180.00 元

C3 建筑立场系列丛书 10:
空间与场所之间
ISBN: 978-7-5611-6650-5
定价: 180.00 元

C3 建筑立场系列丛书 11:
文化与公共建筑
ISBN: 978-7-5611-6746-5
定价: 160.00 元

C3 建筑立场系列丛书 12:
城市扩建的四种手法
ISBN: 978-7-5611-6776-2
定价: 180.00 元

C3 建筑立场系列丛书 13:
复杂性与装饰风格的回归
ISBN: 978-7-5611-6828-8
定价: 180.00 元

C3 建筑立场系列丛书 14:
企业形象的建筑表达
ISBN: 978-7-5611-6829-5
定价: 180.00 元

C3 建筑立场系列丛书 15:
图书馆的变迁
ISBN: 978-7-5611-6905-6
定价: 180.00 元

C3 建筑立场系列丛书 16:
亲地建筑
ISBN: 978-7-5611-6924-7
定价: 180.00 元

C3 建筑立场系列丛书 17:
旧厂房的空间蜕变
ISBN: 978-7-5611-7093-9
定价: 180.00 元

C3 建筑立场系列丛书 18:
混凝土语言
ISBN: 978-7-5611-7136-3
定价: 228.00 元

C3 建筑立场系列丛书 19:
建筑入景
ISBN: 978-7-5611-7306-0
定价: 228.00 元

C3 建筑立场系列丛书 20:
新医疗建筑
ISBN: 978-7-5611-7328-2
定价: 228.00 元

   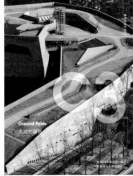 

| C3 建筑立场系列丛书 21：
内在丰富性建筑
ISBN: 978-7-5611-7444-9
定价：228.00 元 | C3 建筑立场系列丛书 22：
建筑谱系传承
ISBN: 978-7-5611-7461-6
定价：228.00 元 | C3 建筑立场系列丛书 23：
伴绿而生的建筑
ISBN: 978-7-5611-7548-4
定价：228.00 元 | C3 建筑立场系列丛书 24：
大地的皱折
ISBN: 978-7-5611-7649-8
定价：228.00 元 | C3 建筑立场系列丛书 25：
在城市中转换
ISBN: 978-7-5611-7737-2
定价：228.00 元 |

  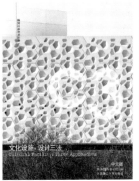  

C3 建筑立场系列丛书 26：
锚固与飞翔——挑出的住居
ISBN: 978-7-5611-7759-4
定价：228.00 元

C3 建筑立场系列丛书 27：
创造性加建：我的学校，我的城市
ISBN: 978-7-5611-7848-5
定价：228.00 元

C3 建筑立场系列丛书 28：
文化设施：设计三法
ISBN: 978-7-5611-7893-5
定价：228.00 元

C3 建筑立场系列丛书 29：
终结的建筑
ISBN: 978-7-5611-8032-7
定价：228.00 元

C3 建筑立场系列丛书 30：
博物馆的变迁
ISBN: 978-7-5611-8226-0
定价：228.00 元

C3 建筑立场系列丛书 31：
微工作·微空间
ISBN: 978-7-5611-8255-0
定价：228.00 元

C3 建筑立场系列丛书 32：
居住的流变
ISBN: 978-7-5611-8328-1
定价：228.00 元

C3 建筑立场系列丛书 33：
本土现代化
ISBN: 978-7-5611-8380-9
定价：228.00 元

C3 建筑立场系列丛书 34：
气候与环境
ISBN: 978-7-5611-8501-8
定价：228.00 元

C3 建筑立场系列丛书 35：
能源与绿色
ISBN: 978-7-5611-8911-5
定价：228.00 元

C3 建筑立场系列丛书 36：
体验与感受：艺术画廊与剧院
ISBN: 978-7-5611-8914-6
定价：228.00 元

C3 建筑立场系列丛书 37：
记忆的住居
ISBN: 978-7-5611-9027-2
定价：228.00 元

C3 建筑立场系列丛书 38：
场地、美学和纪念性建筑
ISBN: 978-7-5611-9095-1
定价：228.00 元

C3 建筑立场系列丛书 39：
殡仪类建筑：在返璞和升华之间
ISBN: 978-7-5611-9110-1
定价：228.00 元

出版社淘宝店

C3建筑立场系列丛书 40：
苏醒的儿童空间
ISBN: 978-7-5611-9182-8
定价：228.00元

C3建筑立场系列丛书 41：
都市与社区
ISBN: 978-7-5611-9365-5
定价：228.00元

C3建筑立场系列丛书 42：
木建筑再生
ISBN: 978-7-5611-9366-2
定价：228.00元

C3建筑立场系列丛书 43：
休闲小筑
ISBN: 978-7-5611-9452-2
定价：228.00元

C3建筑立场系列丛书 44：
节能与可持续性
ISBN: 978-7-5611-9542-0
定价：228.00元

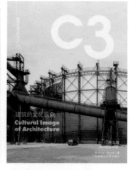
C3建筑立场系列丛书 45：
建筑的文化意象
ISBN: 978-7-5611-9576-5
定价：228.00元

C3建筑立场系列丛书 46：
重塑建筑的地域性
ISBN: 978-7-5611-9638-0
定价：228.00元

C3建筑立场系列丛书 47：
传统与现代
ISBN: 978-7-5611-9723-3
定价：228.00元

C3建筑立场系列丛书 48：
博物馆：空间体验
ISBN: 978-7-5611-9737-0
定价：228.00元

C3建筑立场系列丛书 49：
社区建筑
ISBN: 978-7-5611-9793-6
定价：228.00元

C3建筑立场系列丛书 50：
林间小筑
ISBN: 978-7-5611-9811-7
定价：228.00元

C3建筑立场系列丛书 51：
景观与建筑
ISBN: 978-7-5611-9884-1
定价：228.00元

C3建筑立场系列丛书 52：
地域文脉与大学建筑
ISBN: 978-7-5611-9885-8
定价：228.00元

C3建筑立场系列丛书 53：
办公室景观
ISBN: 978-7-5685-0134-7
定价：228.00元

C3建筑立场系列丛书 54：
城市复兴中的生活设施
ISBN: 978-7-5685-0340-2
定价：228.00元

C3建筑立场系列丛书 55：
灰色建筑中的绿色自然
ISBN: 978-7-5685-0406-5
定价：228.00元

C3建筑立场系列丛书 56：
从教育角度看幼儿园建筑
ISBN: 978-7-5685-0410-2
定价：228.00元

C3建筑立场系列丛书 57：
能源意识与可持续公共空间
ISBN: 978-7-5685-0409-6
定价：228.00元

C3建筑立场系列丛书 58：
灵活的学习空间
ISBN: 978-7-5685-0439-3
定价：228.00元

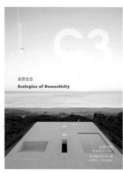
C3建筑立场系列丛书 59：
家居生态
ISBN: 978-7-5685-0455-3
定价：228.00元

C3建筑立场系列丛书 60：
地方性与全球多样性
ISBN: 978-7-5685-0454-6
定价：228.00元

C3建筑立场系列丛书 61：
时间：空间记忆
ISBN: 978-7-5685-0546-8
定价：228.00元

C3建筑立场系列丛书 62：
叩问自然之灵
ISBN: 978-7-5685-0584-0
定价：228.00元

C3建筑立场系列丛书 63：
大学建筑：华丽的转变
ISBN: 978-7-5685-0578-9
定价：228.00元

上架建议：建筑设计
ISBN 978-7-5685-0602-1
定价：228.00元

出版社淘宝店

"C3建筑立场系列丛书"已由大连理工大学出版社出版，欢迎订购！

◆ 编辑部咨询电话：许老师/0411-84708405
◆ 发行部订购电话：王老师/0411-84708943